Time in World History

In this book, Peter Stearns presents the fascinating concept of time through a global historical lens. Covering both calendrical time and clock time, the volume shows how significant changes in conceptions of time are in world history, as they translate many key historical developments from religion to industrialization, into daily experience.

The book explores why and how early societies became interested in measuring time, as well as explaining the causes and ongoing consequences of the modern sense of time. The author compares different societies and cultures in their attitudes and approaches to time and describes the role of globalization in its development. The volume offers many examples and illustrations to aid readers in their understanding of the advantages and disadvantages of various constructions of time, both in the past and among different groups of people today.

Time in World History will be of interest to students of world history and sociology, introducing readers to historical forces that continue to shape their lives quite directly.

Peter N. Stearns is University Professor of History at George Mason University. He has written widely on issues in world history and regularly teaches courses on various facets of globalization.

Themes in World History
Series editor: Peter N. Stearns

The *Themes in World History* series offers focused treatment of a range of human experiences and institutions in the world history context. The purpose is to provide serious, if brief, discussions of important topics as additions to textbook coverage and document collections. The treatments will allow students to probe particular facets of the human story in greater depth than textbook coverage allows, and to gain a fuller sense of historians' analytical methods and debates in the process. Each topic is handled over time – allowing discussions of changes and continuities. Each topic is assessed in terms of a range of different societies and religions – allowing comparisons of relevant similarities and differences. Each book in the series helps readers deal with world history in action, evaluating global contexts as they work through some of the key components of human society and human life.

Aging in World History
David G. Troyansky

The Industrial Turn in World History
Peter N. Stearns

Tolerance in World History
Peter N. Stearns

Gender in World History
Peter N. Stearns

Neutrality in World History
Leos Müller

Globalization in World History
Peter N. Stearns

Time in World History
Peter N. Stearns

Time in World History

Peter N. Stearns

Routledge
Taylor & Francis Group
NEW YORK AND LONDON

First published 2020
by Routledge
52 Vanderbilt Avenue, New York, NY 10017

and by Routledge
2 Park Square, Milton Park, Abingdon, Oxon, OX14 4RN

Routledge is an imprint of the Taylor & Francis Group, an informa business

Library of Congress Cataloging-in-Publication Data
Names: Stearns, Peter N., author.
Title: Time in world history / Peter N. Stearns.
Description: New York, NY : Routledge, 2020. |
Includes bibliographical references.
Identifiers: LCCN 2019042116 (print) | LCCN 2019042117 (ebook) |
ISBN 9780367434946 (hbk) | ISBN 9780367433215 (pbk) |
ISBN 9781003003656 (ebk) Subjects: LCSH: Time–Social aspects. |
Time–History. | Calendar–History. | History–Chronology.
Classification: LCC HM656 .S74 2020 (print) |
LCC HM656 (ebook) | DDC 304.2/3709–dc23
LC record available at https://lccn.loc.gov/2019042116
LC ebook record available at https://lccn.loc.gov/2019042117

ISBN: 978-0-367-43494-6 (hbk)
ISBN: 978-0-367-43321-5 (pbk)
ISBN: 978-1-003-00365-6 (ebk)

Typeset in Times
by Newgen Publishing UK

For Ferris Raymond Stearns Brooks – just in time

Contents

Acknowledgements

A number of people have helped with this study. A big thanks to Kimberley Smith, at Routledge, who suggested the project in the first place and made me wonder why I had not thought of it; and who supported the book as it progressed. Alexis Frambes helped with the research and manuscript preparation, for which I am really grateful. Several colleagues in history and psychology pointed me toward important existing work: Benedict Carton, Robert Houle, Deborah Stearns, Darrin Campen, Yulia Chentsova, Noralee Frankel, Megan Bell. Again, thanks. Robert Ehrlich and Jack Censer read the manuscript in very timely fashion, with valuable comments; and I also thank the readers of the book proposal. And one final note: I get very excited about a project like this, where I learn a lot of new things and try to assemble them intelligently, and this means I burden my family with some of my enthusiasm. So thanks for patience to a number of my children and grandchildren, and particularly to my wife Donna Kidd.

1 Introduction

Time and History

The woman, a fashion designer, has lived in the Netherlands for a number of years. Originally from India, like her husband and children, she now regards herself as half-Dutch, having adapted, among other things, to language and cooking. When asked (in 2019) what adjustment had been most difficult, she immediately pointed to time. When the Dutch say we'll meet at 3 o'clock, they mean 3 o'clock. In India, in contrast, "arriving half an hour late for an appointment is totally acceptable, especially when using traffic as an excuse". She also noted how rare it was for Dutch friends to show up unscheduled, whereas in India it was quite common to say "I was just in the neighborhood and thought I'd stop by." This casualness was another violation of the Dutch sense that life had to be carefully timed and scheduled.

Time is a complex phenomenon. It has its scientific side – various aspects of time have been research subjects at least since the early civilizations. It obviously has a technology side, in the vast array of devices that have been invented, and continue to be invented, to keep track of time. But it also has its very human side, as the immigrant story suggests, conveying much about how people organize their lives and their interactions with others.

Most important, for world history, is the fact that time changes. Modern society – even in urban India, despite a distinctive culture, is deeply time-conscious. One of the implicit functions of school is to drill a sense of clock time into children, and many will carry this lesson throughout their lives. But this pattern did not always prevail, which is where the history comes in. Figuring out how the meanings and measurements of time have shifted says a great deal about how peoples' lives have changed – and this gives us some perspectives, as well, on some of the downsides of time-consciousness in contemporary life.

All of this may come as a bit of surprise: working on a history of time often elicits a response, how can there be a history of that? Time can seem natural (and some of it does link to nature) or preordained. But, in fact, all sorts of times vary depending on the region and historical context, and all sorts of times, and more important all sorts of human experiences with time, have changed, sometimes at many points since human beings first began to look up at the skies to try to figure out what was going on.

Why do we need to know what time it is in the first place? And when did people start trying to answer the question?

There are of course many kinds of time, and a time-focused history will touch on a number of them. Seasonal time is a pretty obvious topic, since it is hard to miss and at the same time helps us anticipate changes in temperature and other conditions: charting seasonal time was one of the first targets for early human societies, even before the advent of agriculture and civilization. Historical time is far more complex, and many groups long lacked what professionals would call a real historical consciousness. Many peasant communities, for example, could blend recollection of a disastrous war or flood that happened a few centuries back, with events that occurred just last year, and they had no sense of the distinction. Personal historical time is another interesting variable. Regular celebrations of birthdays did not begin to occur, for ordinary folk, until the middle of the 19th century and, even today, in many parts of the world, lots of people do not know the date or even the precise year in which they were born – again, suggesting another difference in the way time can be perceived.

This book will deal, at least briefly, with all these time frameworks, with additional attention to the time unit that is one of the purest human inventions, with no connection to natural patterns, the introduction of the week. Ways to calculate annual calendars play an important role both in the history of time and in human life, including religious life, and the ongoing history of calendars – involving some issues even today – deserves careful attention. But the main focus will be on what, at least until the recent advent of cellphones, would have been called clock time – the kind of time around which individuals and groups may organize much of their day. We will be asking when, how but also why people began to wonder about the time of day, and then how this issue has changed from one world history period to the next. The result reveals a great deal about basic patterns of life.

The inquiry will involve some of the big hitter topics in the human experience. Time often relates closely to religion – think of calls to prayer or church bells. It certainly links to how people work and organize work. It links to politics; we will see how important time was to officials in the courts of the Roman Empire, as they bustled, much as modern lawyers do, to show up promptly. The capacity to rearrange time, or try to, has played an important role in statecraft in many periods. Time can be an essential feature of trade, not only in the sense of when stores open but also in the centrality of time calculations in navigation; time and the spread of oceanic commerce become integrally connected. More gradually, time would also come to embrace leisure: one of the big innovations, in the modernization of sports in the 19th century, was the introduction of the notion that games should be timed (and some older sports, like baseball, are still not timed even today, which can pose a bit of a problem in a time-obsessed society).

Again, among all these big connections, it is important not to lose the human or personal side of the story. When did being punctual begin to matter, and in what aspects of individual and group life, and where does it matter today? Have modern societies, at least in the West, overdone time-consciousness, and how do people, consciously or unconsciously, give themselves some respite from timed lives?

Several vantage points come into play in dealing with time. There is the science side. Time became a fundamental aspect of research in physics in the 20th century. For many centuries, from Aristotle through Newton in the Western scientific tradition, time was assumed to be an absolute: that is, if your clock was mechanically correct, it would tell the same time no matter where it was placed. But discoveries from the late 19th century onward, often linked to new calculations in astronomy, made time a relative concept, and the relativity of time in turn became a fundamental feature of Einstein's revolution in the field. Time moves faster the farther away one is from the surface of the earth – faster, thus, on a mountain than at sea level; this means that a minute or a second has a slightly different duration depending on altitude. These vital embellishments on the scientific approach to time obviously provide their own illustration of the fact that ideas about time change from one historical period to the next.

Time has long had a technology angle, often related to the science involved. There is of course no need for a special device to determine the basic difference between night and day – though in several religions calculating the precise time of sundown can demand attention. But as soon as people became interested in more detailed divisions of the day, they faced an obvious problem: it is really hard to use natural observation to tell what time it is at night, or even on a cloudy day. So, as we will see, artificial devices began to be devised fairly early in the human experience, at least once the complex societies we called civilizations emerged. Learning about advances in timing devices in turn became one of the leading components in the contacts among different societies; the Chinese, for example, were often eager to import better instruments from societies they were connected to by trade. And the story of technological improvement in timing devices obviously not only continues but intensifies. Today national standards offices in many countries maintain a network of interconnected atomic clocks (based on microwave or electron transition frequencies) that provide unprecedented accuracy in timekeeping: currently this system will lose less than a second in 300 million years – though scientists are eager to improve even on this performance. And, for accuracy junkies, it has been possible to buy atomic clocks for home or office, for the past several decades.

More important than science or even technology, time also involves the human element, deeply embedded in the ways that people and organizations function. Here is where changes and contrasts are most interesting and meaningful. When and why did it become important to show up for work "on time"? When did it become desirable (and technologically feasible) to wake up at a particular time (Hint: the first alarm clock was invented in the late 18th century, though there are some antecedents even before mechanical clocks were introduced)? When and why (as well as how) did it begin to seem essential not just to determine who won a race, but to time the process?

Or take the phenomenon of visits to one's home, a subject already evoked by the Indian immigrant in the Netherlands. When friends or associates visited George Washington in Mt. Vernon back in the 18th century, they had to come from some distance and for that reason would usually stay at least for several days: this was simply a function of the nature of travel. But it would have been impossible to predict when they would arrive (and often, probably, when to encourage them to leave). A letter might establish what day it would be, but greater precision would have been very difficult. For many contemporary Americans, this degree of vagueness in the timing of a guest's arrival would be almost intolerable; our social lives are organized around a very different sense of personal time.

And this personal aspect of time becomes wrapped up in all sorts of additional features. Punctuality or tardiness came to serve as measurements of essentially moral qualities, certainly relevant to modern job performance. Degrees of patience or impatience are intimately connected to time. How tolerable is it for a game or a show to start "late"? When should schooldays start? Here, current debates about time bump organizational convenience against the presumably peculiar features of the teenagers' body clocks, with growing pressure to move start times back a bit. And college students become far more reluctant to take "early morning" classes than their counterparts were a few decades back, another link between change and variety in time and personal experience.

Ultimately, the history of time links science, technology, social needs, and personal reactions to tell the story of a really important, if underappreciated, aspect of the human experience. Time provides a surprisingly sweeping yet intimate way to evaluate a number of features of world history, and to connect them to ways people live today.

The focus, as already suggested, will be on change, though sometimes very gradual change – or putting the same point another way, how different the past is from the present, where time is concerned. Though to many contemporaries time would not at first blush seem to be a historical variable – after all, an hour is an hour – in fact even our notion of the hour is a fairly recent though not brand new invention. Exploring the development and evolution of key concepts and articulations of time, particularly into modern preoccupations, offers really important insights into some fundamental shifts in the human experience.

But the emphasis on change must be complicated by several additional factors. First, as in virtually any case where change is involved, there are also continuities from the past. Thus even amid the modern pressures of time, people continue to seek some areas where an older, more relaxed time sense can prevail. This was one of the motivations behind the introduction of vacations as a normal part of the annual experience, and it also defines the growing effort, in the later 19th century, to establish a two-day weekend. Of course, it turns out that time-consciousness can intrude even on vacations and weekends – a major issue in the age of the Internet –, but for some people at least some real continuities with earlier definitions of time are not entirely erased.

Changes in time, even ones that today seem inevitable, also provoke protest and debate, precisely because time reaches so deeply into personal experience and expectation. The timed rhythms of the factory floor antagonized many workers, particularly in the early decades of industrialization, who sought both personal and collective ways to slow things up. Efforts in the later 19th century to redefine time zones, in the interests of global coordination, provoked lots of concern. Disputes over time standards were also wrapped up in reactions – sometimes, violent reactions – to imperialism. Today, debates over whether to preserve a distinction between standard and daylight savings time (occurring in several parts of the world, and not just the United States), can elicit a surprising amount of passion.

Changes in time's history also occur at a different pace, depending on the region and group involved. The chronology of the relationship between redefinitions of time and industrialization obviously varied with the dates of industrialization itself – still an issue today, from a global standpoint. But variations in chronology are not just a modern phenomenon. Different societies developed ultimately common mechanisms to measure time – like the clock – at different dates. Even within a single society, different groups may display a distinctive pace. Into the 20th century, for example, many people believed that men (involved in workplace time) and many women (involved in the very different timing of housework) had different responses to the modern apparatus of time, though ultimately – by the later 20th century – these would increasingly merge. So, even where change ultimately becomes global, the variations in chronology require attention.

And this bleeds into one of the most complex topics of all: the extent to which different groups and societies maintain somewhat different cultures of time, even as they participate in some common changes. The different conceptions of time suggested by the Indian immigrant to the Netherlands may in part reflect distinctions in timing – the Dutch have been exposed to modern time systems longer than many Indians. But it may also reflect more durable cultural values, that will continue to carve out different time experiences well into the future. Any world history of time thus must allow for the importance of *comparison*, bringing the subject of time into one of the standard analytical categories of the field.

Cultural differences over time, however, whether real or imagined, have sometimes been more than analytical categories, but also sources of serious bitterness and prejudice. Apparent failure to live by a rigorous modern time sense has often been taken as a sign of laziness and inferiority, another factor that persists today in a number of instances. Here is a tricky element to take into account – carefully – in dealing with time as a major facet of world history.

Finally, and this is a standard issue in dealing with powerful patterns of change, it is important not to equate change and progress. Most modern people would be very uncomfortable if they had to go back to societies defined by earlier ideas about time and earlier technologies for measuring time. It will be easy to assume that developments like clocks and scheduling represent clear advances over prior ignorance and imprecision. But the historical evolution of time has brought cost as well as benefit. One of the advantages of a historical approach to time is in fact the opportunity to stand a bit apart, and to debate the minuses as well as plusses of time systems that most of us view as normal.

A global history of anything, particularly a brief one like this book, inevitably raises important issues, beginning with available evidence and relevant research. The history of time has not been studied as extensively as one might imagine, given the importance of the topic to human life. And, as in so many world history fields, there is a great deal of variation in regional adequacy.

Happily, there are some terrific individual studies, on which the following chapters rely extensively and which are noted in the "Further Reading" sections. The early modern West has been examined particularly extensively, with some interesting disagreements in the leading studies. The history of clocks, though sometimes a bit antiquarian, also extends back chronologically, particularly for the Middle East, Europe and China. And there is a good bit of important work on the development of calendars, in several civilizations including pre-Columbian

Latin America. More modern history (both calendars and clocks), ironically, has drawn less attention, but there is excellent scholarship on time and modern work; and on Africa; and on aspects of time and popular resistance in globalization. On the other hand, the history of time in India has drawn less attention until the colonial period and, while there have been appeals for more attention to the subject in Latin America, many opportunities remain. And even for the West (Europe and the United States) there are some time-related subjects that could generate more historical attention: we will refer for example at several points to the possibility of a history of patience and its relationships to changes in time-consciousness, but there is little work available.

Again, as with most world history, the history of time raises standard questions about comparative differences and similarities. Different regions and cultures have different histories, and the kind of analysis that results will inform many of the chapters that follow. But two cautions: first, not everything is variable, if only because time relates both to human nature and to nature, and these impose some standard patterns. Similarities as well as distinctions are important, including of course some widely shared modern trends in response to common developments like train rides or the increasing availability of watches (and, recently, smartphones). Second caution: as suggested above, time patterns often rouse deep prejudices, with groups who maintain a distinctive time sense held as morally inferior. This is not a useful historical or contemporary judgement, but it must be kept in mind in any comparative work on time.

The following chapters cover basic chronological periods in the history of time, often closely linked to some of the standard basic periods in world history.

Chapter 2 deals with time in hunting and gathering societies, and some of the first efforts to measure at least certain forms of time. Agriculture introduced new needs and new opportunities to deal with time. The advent of civilization brought further change. Scientific study of time increased and major new devices were introduced. More complex religions required new uses of time for the sake of regular and coordinated ceremonies. People also began to feel a need to time certain activities – like public speeches – and this required some new technology. Chapter 2 thus treats some of the basic issues in the emergence of increasing awareness of time and of the means to implement this awareness. It also highlights some basic divisions – most obviously between urban systems and the rural majority – that long persisted in time-consciousness.

Chapter 3 focuses on the great civilizations, in the classical period but also the post-classical centuries. The great religions further expanded the importance of timed activities – the Islamic call to prayer at set intervals in the day was the most obvious example. Further innovations in measuring devices reflected changes in science and technology alike. Other systems, like expanding law courts, compelled new attention to time and scheduling. Increasing contact among civilizations through trade also encouraged exchanges of innovations, as in the Chinese interest in timing devices developed first in Persia. This long period did not see revolutionary developments in the history of time, and many routine activities were still essentially untimed, but there were important new interests and needs.

Chapter 4 centers on new developments in the early modern centuries (1500–1800), which turns out to be a crucial period in the history of time. Steady and often dramatic changes in the technology of clocks highlight these centuries, including the first invention of devices like watches and alarm clocks. The important relationship between time and navigation reflected major change as well. But there was also a growing cultural interest in time and punctuality: the 18th century would see the most active discussion of these topics not only compared to the past, but also to the following period. Beyond technology, though closely linked to it, some fundamental changes were occurring in the ways some people thought about time, making this perhaps the most challenging period of all in terms of assessing the nature of change and its causes. But there is another issue as well: the leading changes were occurring in Western Europe. A leading scholar has argued that Western innovations about time were fundamental to other changes in Western society, ultimately including industrialization. But Western innovations had some impact on other parts of the world, an important comparative topic in its own right. Time provides a significant opportunity to tackle the thorny question of Western "leadership" in a basic category of human and social organization.

Chapter 5 centers on time in the industrial revolution of the West, covering the 19th and early 20th centuries as industrial systems developed and generalized. Here is where new ideas and devices began most obviously to transform daily life and consciousness – particularly at work, but also in leisure and childrearing. It became increasingly important to have timepieces in the home and on the wrist, though sometimes this was more for show than for active use. Resistance to modern timing emerged as well, along with new efforts to provide occasions for the relaxation of time pressures including the two-day weekend (initially called, in some places, the "English week").

Chapter 6 covers much the same chronological period – 19th and early 20th centuries – but with a global focus. Several related questions define the chapter: how would the new, "industrial" time sense affect interactions with largely agricultural societies, for example in growing world trade? How was the spread of imperialism affected by cultural differences in time sense – including of course the emergence of deep-seated time-based prejudices? But the chapter also covers the emergence of new global time systems, reflecting the needs and power of industrial societies, including the division of the world into time zones in the later 19th century.

Chapter 7 deals with the past hundred years, mainly around the intensification and globalization of the modern requirements for time. New communications and transportation devices – including jet aircraft – introduced vital changes in the experience of time; the new term, jet lag, introduced in the 1960s reflected important shifts in the human experience of time. The advent of time and motion studies at work reflected a desire to override personal foibles in insisting on adherence to scheduled time. Modern time-consciousness, and associated devices, spread increasingly to most parts of the world, though particularly in the growing cities. Childhood became even more time-defined, with major implications for the nature of play. At the same time, some of the psychological downsides of time pressure also emerged – the new concept of stress or burnout was a time-related ailment, identified as a major issue by the World Health Organization as recently as 2019. And there were new efforts to develop alternatives to modern time pressure; this was an important if largely implicit focus of the counterculture movement of the 1960s and aspects of the positive psychology movement that emerged after 1998. The contemporary period of world history is marked most obviously by the increasing rigor of modern time, but it also sparked new kinds of resistance as well. Finally, comparative differentiations in the specifics of time continue to loom large even as, at the scientific and political levels, coordination around the importance of precision steadily increased.

Finally a concluding chapter returns to some of the complexities in the history of time, including time and the contemporary life cycle, with some stages in life more affected by rigorous timing than others.

The history of time uncovers a host of important and revealing features, centered on some of the most basic aspects of the human experience. The emphasis on significant change is obvious. So, from early human society to the present, is an intriguing balance between shifting *needs* – around work, religion, the organization of the state – and *curiosity* – the desire to contemplate how to measure time simply for its own sake. Sometimes, needs seem to predominate, as in the timing rules of modern schools and factories. But sometimes changes in the assessment of time, including major inventions, primarily reflect sheer inquiry, not closely tied to economic demands. Figuring out how this combination operates, in each of the major world history periods, helps explain why, in human terms, time has not stood still.

Further Reading

Several histories, mostly focused on technologies, are useful on the evolution of time and timing:

Barnett, Jo Ellen. *Time's Pendulum: The Quest to Capture Time – From Sundials to Atomic Clocks*. (New York: Plenum Trade, 1998).

Jespersen, James, and Jane Fitz-Randolph. *From Sundials to Atomic Clocks: Understanding Time and Frequency, Second Revised Edition*. Second Edition, Revised edition. (Mineola, NY: Dover Publications, 2011).

Landes, David S. *Revolution in Time: Clocks and the Making of the Modern World*. (New York: Barnes & Noble Books, 1998) – one of the most challenging books on the subject but mainly early modern in focus.

Levy, Joel. *Really Useful: The Origins of Everyday Things*. (Collingdale, PA: Diane Pub Co, 2002).

Milham, Willis Isbister. *Time & Timekeepers: Including the History, Construction, Care, and Accuracy of Clocks and Watches*. (London: Macmillan, 1923).

Richards, E. G. *Mapping Time: The Calendar and Its History*. Revised ed. edition. (Oxford: Oxford University Press, 2000).

Waugh, Alexander. *Time: Its Origin, Its Enigma, Its History*. 1st Carroll & Graf Ed edition. (New York: Basic Books, 2000) – somewhat erratic and at points problematic, but with interesting insights.

2 Why Time? The Experience of Early Societies

An anthropologist, studying the Nuer people of the Sudan (Upper Nile River), noted their distinctive and very pragmatic approach to time. Their region alternated between wet and dry seasons; during the latter, they had to fan out to camps in search of food, whereas during the rainy period they could settle in a village. Their division of time highlighted this seasonal distinction, and they named the two main seasons by the activity involved, like going to the camps. They had no weeks, presumably because they did not need market days, and they had no idea of hours of the day. When referring to an event in the recent past, they counted by the number of sleeps they had enjoyed; otherwise, for a longer period, they relied on phases of the moon. One historian has dubbed their approach "eco-flextime".

Obviously this is a sense of time very different from our own, and different as well from other societies, even fairly "primitive" ones, that saw a greater link between time, the gods, and religious ritual. It suggests both the need for some concept of time that all groups probably experience, if only to explain changes imposed by nature on daily life, and the varied bases on which this time concept can rest.

This chapter covers early human societies through the advent of agriculture and the first, river-valley civilizations – a huge chronological span. The focus is two-fold: most obviously, we will discuss the emergence of new studies of time and new instruments to measure it as more organized human societies developed. But we will also be exploring the more difficult question of why people became interested in time and how, if at all, new ideas and devices affected their lives. Linked to both these issues is the nagging question of evidence: how do we know anything about the early human experience of time and what major gaps remain?

For hundreds of thousands of years, human societies were centered on a hunting and gathering economy, grouped in small, mobile bands that usually numbered about 40 to 60 people (about half of them children). These groups, often successful, possibly enjoying greater satisfaction than people do today in part because of their harmony with nature, had little need for time, particularly in terms of daily measurements. This is the starting point in the rich history of time, and it allows for some important exceptions and uncertainties. In the main, however, hunters and gatherers did not live rigorously timed lives and, apart from obvious natural markers like night and day or the shifting seasons, many of them probably had little awareness of time in any form.

It is also important to remember that humans have a natural circadian or bodily rhythm that coincides pretty closely with a day – the correspondence is not exact, since the cycle lasts 11 minutes over 24 hours. But this means that with no artificial devices or even exposure to light, people have a cycle that will include fairly regular timing for wakefulness, hunger and sleep (though it can be overridden by other factors, such as, in contemporary societies, night work or jet travel).

Two factors limited the need for any detailed idea of time beyond what could be experienced from body rhythm or observed from the rising and setting of the sun plus, during the day, the visual evidence of the sun's position in the sky. The small size of the basic groupings made it easy to coordinate by voice and contact, without any need for formal scheduling. A group could simply launch its day, usually after sunrise, by assembling members directly. If time was a factor – some hunting, for example, was best done early or late in the day, simple observation and small-group communication – "shall we get going now?" – would suffice.

Added to this was the fact that most hunting and gathering groups did not spend much daily time at work. There is some debate here, but most assessments of the average work commitment range from under three to about six hours per day. This obviously left a lot of time for unscheduled activities and put no particular pressure on a need to get the group together (whether for hunting or for gathering) with any urgency. Surely there were exceptions to this latitude, when weather conditions or some other problem made game scarcer, in turn requiring more effort. But there is little question that, ordinarily, hunting and gathering people lived with little time pressure. This meant in turn that some of the earliest efforts to go beyond simple observation in charting time may have arisen more out of curiosity than any particular need.

One of the earliest available signs of an interest in time dates back over 20,000 years, in a bone discovered in a valley of the present-day Democratic Republic of Congo. The bone is carefully carved with hash marks, which may have represented an effort to count days. Obviously this interpretation is uncertain, but the carver might have been trying to keep track of the length of some trip or expedition that went beyond the ordinary. The effort, of course, did not involve any attempt to measure time *during* a day or project time into the future.

Seasons and Measuring by the Moon

A great deal of the early interest in time centered on charting the seasons and even dividing the year into something like months. Here too, for many people, simple observation might suffice: it was usually clear enough when summer weather was fading or winter losing its grip. But weather, obviously, is not entirely predictable, and some people, particularly inquisitive or mathematically minded, might have wanted more certain evidence of what time of year it was, possibly even in advance of the seasonal turn.

At least three factors could prompt this kind of interest. From a practical standpoint, particularly in some climates, it would really be helpful to know when a season like summer normally ended, because this would mark changes in the availability of migratory animals or the seeds, nuts, and berries sought by the gatherers. Knowing what was normal, in terms of anticipating a schedule beyond the vagaries of daily weather, would help planning. It might be useful, in other words, to come up with some kind of calendar. We have seen this kind of interest among the Nuer people.

Second: while most hunting and gathering groups were small enough to require only personal coordination, most early societies also encouraged periodic assemblages among several bands in a given region. These meetings were desirable to help keep the peace – hunters-gatherers were eager to avoid war: gifts could be exchanged, and common feasting would creative positive bonds. Further, group meetings also served to promote intermarriage, desirable both for peacekeeping and to avoid too much inbreeding within a single small group. So: it would surely be useful to have some agreed-upon seasonal points at which modest assemblies could occur. Further, identifying the timing of a season when local food resources could support a larger grouping – like the beginning of summer – provided another motivation for seeking some degree of accuracy as well as coordination.

And third: there might be religious and ritual reasons to agree on some particularly important days during the year. Some groups, eager to propitiate the sun, developed ceremonies to note when this body was at its high point, the summer solstice, or at its lowest ebb. Larger-than-usual gatherings might occur to join in celebration or propitiation. In some cases, particularly large burial sites might be located in conjunction with the assembly points, joining seasonal worship to the honoring of ancestors.

All of this means that the earliest and most elaborate evidence we have about human interest in going beyond simple observation in the measurement of time involves attempts to pinpoint solstices or to record the cycles of the moon or to chart configurations of the stars – or some combination of these techniques. Much relevant research may have occurred early in the human experience, or at least the experience of Homo sapiens once the species gained the power of speech, and left no record. But the evidence that is available is intriguing.

Thus several groups in Africa, such as the Borana people (present-day Kenya and Ethiopia), ultimately developed a sense of annual calendar by observing the position and brightness of star clusters in the Orion constellation. Orion contains some of the brightest stars visible during winter, and could thus be used to help chart the seasons.

An arrangement of stones near Victoria, Australia, dating back at least 11,000 years, was probably designed to keep track of the solar year, while several ceramic objects found in the Balkans also seem to represent early efforts at keeping a calendar, based on movements of the stars.

But calculations based on phases of the moon may have been even more common. Warren Field, in Scotland, is the site of what some claim is the world's oldest known calendar, built around 8000 BCE, and first rediscovered, from the air, only a few decades ago. (Several similar sites, a bit more recent and less elaborate, have also been found in Scandinavia.) The Warren Field calendar consists of 12 moon-shaped pits – one, presumably, for each lunar cycle in the year; the pits were of different sizes, probably to designate clearly the 12 separate cycles during the lunar year. But the basic location was selected because it allowed an accurate sighting of the mid-winter solstice. Collectively, the sequence of pits presumably designated and forecast the cycles to come over the ensuing months. There is much earlier hunter-gatherer evidence, in France for example, that some people were noting moonrises through objects like carved bones, but these were simply records of what had occurred and did not anticipate time for the future, which is after all the clearest innovation involved in creating a calendar. Warren Fields suggests that simple observation of the skies was now being translated into seasonal predictions, indicating the future phases of the moon to pinpoint more precise dates. Archeologists cannot say, of course, whether the goal was better seasonal information for hunters (in a northerly location where animal migration could be significant) or to help organize larger-than-usual rituals, or perhaps both. Certainly the group effort involved, in digging the sites and hauling in some large stones from a distant location, indicates considerable investment.

Keeping track of the moon's patterns – from one full moon to the next – does permit the possibility of dividing a year into 12 segments, which is where simple observation could be built into more extensive measurements and forecasts. And lunar calendars continue to loom large in the contemporary world, most obviously in Islamic rituals such as the timing of the fasting month of Ramadan.

From an accuracy standpoint, however, purely lunar observation, multiplied by 12, falls 11 days short of what we call a year based on the migration of earth around the sun, for the average lunar cycle is only about 29 days. So many early societies learned to supplement lunar observations by annual corrections derived from the earth's rotation around the sun – or, once more, an annual solstice. The Warren Field site, with its opportunity to identify the solstice, may have built this corrective in directly – hence the linkage of 12 moon cycles to sighting the seasonal position of the sun. But the importance of combining solar observation with the more natural calculations about the moon led to some even more famous early efforts to measure seasonal time accurately, such as the monumental grouping of stones in the British site known as Stonehenge.

Stonehenge, now one of the most iconic tourist sites anywhere in the world, began to be constructed in southern England around 3000 BCE (or possibly earlier). Initially it seems to have been a special burial ground, and this usage persisted even as other functions were added. Markings for a lunar calendar, similar to Warren Fields, may have been constructed as well. Later building phases featured the importation of huge stones, many of them from a considerable distance, to add to the religious functions of the site and, most important for our purposes, to help record the annual positions of the sun. For one line of stones, supplemented by a circle of timber, was carefully aligned to pinpoint the setting of the sun at the summer solstice. Another line of stones highlighted the sunset of the winter solstice but also sunrise for the summer event. Ritual services may have focused primarily on the winter – that is when, judging by surviving animal bones, the largest human groups gathered, possibly to offer prayers and sacrifices designed to assure that the sun would gain strength again after this annual low point. Some scholars believe that the stones and their ritual functions were also seen to have healing properties. For our purposes, the main points are, first, the huge interest in marking the positions of the sun and the ability to do so with great accuracy, and, second, the linkage of this interest to religious rituals and ceremonial gatherings – some of which drew people from other regions. Marking seasonal time could clearly warrant substantial investment and generate a wide audience.

Evidence of this sort can be multiplied by findings from many other regions; there are numerous other sun-sighting stone groupings in Britain but also other parts of northern Europe, though none is as ambitious as Stonehenge. Somewhat later, groups in the Andes in South America – antecedents of the Incas – constructed stone piles or pillars for sighting both summer and winter solstices. Constructions of this sort clearly demonstrate the interest and capacity of key hunting and gathering groups to move from intelligent observations of sun, moon and seasons to much more elaborate and accurate calculations. Conventionally, the notion of calendars and formal mathematical calculations of divisions of time was long associated with the rise of civilizations in places like Egypt and Mesopotamia, and indeed there are important new contributions to note. But the interest in time measurement, and the capacity to invest heavily in estimating and celebrating, goes back into much more primitive societies.

But while these achievements might impact ordinary people significantly, at least in some periodic rituals and seasonal planning, they had nothing to do with daily life or routines. Neither necessity nor curiosity generated major innovations at this level until considerably later. At most, there is scattered evidence, from late in the hunting and gathering period, that some groups may have put up rods or sticks to cast shadows from the sun, to help determine its stage during the day. But we do not know how widely this was done, or for what purpose. Major developments in this category would have to await the advent of agriculture and civilization, which generated both new needs and new capacities.

Some hunting and gathering groups did, however, begin to introduce one additional time division, that sat between monthly calculations and potential divisions within a single day: in a number of regions: they pioneered in developing the idea of the week, which would long stand as an important marker and would indeed be further embellished in reaction to other changes in the use of time in modern societies.

The Week

Like so many features of the hunting and gathering phase, the origins of the idea of a week are unclear, as is the rationale – for, again, the unit has no necessary connection with phenomena in nature. The purpose of the week is simply to introduce a special time in which ordinary daily activities may be modified – most commonly, to allow for special religious functions. Some designations of the week may simply have expressed a desire to devote each day to a major god, in a recurring cycle – without any special commitment to a weekend ceremony.

As with so many aspects of early human history, it is not certain when various peoples began trying to designate weeks. In some cases, the process seems to have begun among well-established hunting and gathering groups – like the Northern Europeans around 500 BCE – but in other cases it may have awaited early agriculture. What is clear is that the impulse to identify weeks was widespread but that the specifics long varied greatly depending on the region involved.

For the most interesting initial feature of the week, aside from its emergence in most regions in the world either through imitation or spontaneous creation, was the fact that it did not necessarily have seven days. This makes sense in some ways: since the unit is artificial, different groups might well conceive of different cycles. And given the huge dispersion of the human population during the hunting and gathering period, specific regions might well come up with their own definition, which would inevitably differ from the inventions of other groups in other places. Finally, a few major regions, particularly in parts of Africa, did not develop the concept of a week at all – which would turn out to matter in more modern times.

Variations, where the week did take hold, were intriguing. Thus the Germanic peoples established a five-day week, with each day named for one of the major gods. (The English language retains some of the nomenclature: thus Thursday or Thor's day, Wednesday or Wotan's day and so on.) Romans long used an eight-day week, though they would ultimately shift to seven days. Ancient Chinese and also Egyptian calendars had ten-day weeks. (Ironically, an effort was made during the French revolution, in the 1790s, to introduce a ten-day week on grounds that it better fit into a rational decimal system, but it was vigorously rejected because it so prolonged the work week.) Both Java, in present-day Indonesia, and Korea initially developed five-day weeks, while the Igbo people, in Nigeria, had a four-day cycle.

The main point, clearly, was a widespread impulse to impose some time measurement on the otherwise monotonous cycle of days during a month or a year. Beyond the simple human impulse to measure or impose some regularities, religious functions surely provided the clearest practical reason to devise a week: with this unit, one day could be clearly set aside for special worship purposes – as would be clearly evident in the introduction of the idea of the Sabbath day in early Judaism.

Once the idea of the week was established – and again, in many regions, this seems to have been accomplished initially by hunting and gathering groups – it might later undergo two modifications, both dependent on the emergence of more complex societies based on agriculture. First, weeks allowed designation of a market day along with a special religious day (the days might or might not overlap). And, second, fuller knowledge of astronomy could encourage a partial linkage of the week with natural phenomena.

Market functions were clear enough, and still define uses of the week in many modern societies. Hunting and gathering groups were largely self-sufficient. They traded irregularly, often focused on exchanging ceremonial gifts that might help keep the peace or provide some special ornamentation. There was no reason to set aside a special day for this kind of function. With agriculture, however, some trade became more essential. Most agricultural communities produced most of their own needs locally. But there was often some food surplus, and also an interest in buying a few products either from the outside world or from local artisans who made some items (like ceramic pots, for food storage) better than an individual household could. And certainly early agricultural cities depended explicitly on trade, to provide necessary foods. At the same time, in rural villages and even many cities, there was not enough business to support regular shops that would operate on a daily basis. Hence the obvious interest in adding to the functions of the week by setting aside a special market day, when vendors might come in from neighboring areas seeking to sell foods or other goods. And while market days leave little trace in contemporary consumer societies like the United States, where stores and now online facilities are open all week long, they persist in many cities in other parts of the world; here, open-air stalls faithfully show up once a week to supplement the more regular commercial outlets. And all this operates alongside the even more traditional use of the week to set aside a special ritual day.

Efforts to attach the week to more systematic thinking about calendar time constituted the second extension of the concept after societies had moved beyond the hunting and gathering phase. In Central America, for example, both the Mayans and Aztecs devised ritual calendars that identified a ritual cycle of 260 days, each divided into 20 weeks of 13 days each. More important, for more practical purposes, they also divided the solar year itself into 18 periods of 20 days each (with five nameless days added in), creating 20-day months, each in turn carved into four five-day weeks. The end of each five-day week was a market day. Here, clearly, was an effort to link the week to a solar cycle and make it more mathematically rational in the process.

But the most important effort to define a week, possibly in relation to natural observation, ultimately took shape in the early civilizations of the Middle East, and would produce the seven-day week that we all know and love today.

The idea of a seven-day week first shows up by the 6th century BCE, where, among other things, it was built into the emerging Jewish religion and its identification of the Sabbath. Why seven days? There are several hypotheses.

Some scholars suggest that the choice of seven was an effort to link weeks with divisions of the lunar cycle – roughly 29 days could best be divided into four seven-day spans. They believe that the calculation was probably devised by priests and astronomers in Mesopotamia and then borrowed by the Jews. They find evidence, for example, that the earliest Mesopotamian civilization, in Sumeria, used a seven-day unit for astrological purposes.

Other interpretations, however, argue for a more distinctive Jewish origin, finding no evidence of this choice in the other societies that neighbored Jewish territory. This would mean, in turn, that the specification of seven days was based primarily on religious needs, not any more elaborate calculation, though there may have been some interest in linking the Sabbath (once every four weeks) to the cyclical recurrence of a full moon.

But once established, it is clear that the seven-day concept proved very influential, spreading widely in the Middle East and then beyond. Persians used a seven-day week, and a seven-day cycle was incorporated into the religion of Zoroastrianism. Greeks had adopted it by the 4th century BCE, presumably from their contacts with the Persians. Alexander the Great's conquests spread the concept into northwestern India. Romans converted to the seven-day system through their own interactions with the Greeks, abandoning their earlier eight-day cycle (by 45 BCE). Then, of course, the unit became fundamental both in Christianity and in Islam, both of which would introduce it to additional regions through their missionary efforts.

The spread of the seven-day week to East Asia is somewhat more mysterious. The first Chinese reference occurs in the 4th century CE, and may have resulted from contacts with the Middle East. It was more firmly established by the 7th century, and then imported into Japan as early as 1007 CE. The unit also spread widely in India.

Overall, the early history of the week, with all its intriguing complexities, suggests two key points central to the study of time in world history more generally. First: humans really do like to carve out regularities in time, even beyond those clearly generated by nature. This helps create a sense of order and, in some cases, organizes links to divine forces. Second: time designations introduced in one region may well prove influential in others, as ideas about time prove readily transportable. The realization that much of the world had accepted a seven-day week well before the apparatus of early globalization is an interesting illustration of the importance of contact. The fact that seven days might also appeal to lunar observers or to astrologers, for whom the number seven had special significance, must be considered as well, in promoting acceptance of this particular definition.

Finally, far more than with lunar months or solar solstices, commitment to a week (of whatever length) clearly impacted the routines of ordinary people – as is still true today. It helped regulate work and worship, and also market opportunities, on a regular basis. But it still left open the question of whether there was any point interjecting time into a single day, beyond the day–night divisions imposed by nature. And here, major innovations would come far more slowly.

The Role of Agriculture and Civilization

We have seen that hunting and gathering peoples seem to have done little to identify the passage of time during a day, though we cannot neglect the scattered evidence of the use of markers to measure the position of the sun in the interval between rising and setting. Efforts to define calendars and weeks advanced more impressively, moving beyond simple observation; but there was no particular technology yet attached. Not surprisingly, the advent of farming and then the emergence of more complex societies or civilizations not only built on the prior achievements but added major innovations as well.

In world history generally, the replacement of hunting and gathering by agriculture forms one of the great divisions in the human experience. The transition was complex: it occurred at different dates in different regions, and was often resisted by hunting and gathering groups. In some areas it was also rivaled by the possibility of creating a nomadic herding economy, where needs and opportunities for considering time probably did not change much from what hunters and gatherers had already established. There is evidence that some nomadic groups kept quite accurate tallies of the major solstices – in one case in the 12th century, one Asian group had better calculations than a representative of the imperial Chinese government, who had to argue that 15 minutes did not make much difference when it came to the winter solstice. But they did not pioneer any really new measurements

But where agriculture did triumph, it brought huge changes in the human condition, often in very personal ways. Residence was transformed. Most agricultural societies created settled communities – villages – replacing the earlier pattern of moving around periodically in search of game. Most peasants stayed put, for the investment in clearing land and digging wells placed a premium on attachment to a defined family property. Many early agricultural communities, furthermore, developed irrigation systems, that provided yet another reason to stay in one place.

Gender relations shifted. Hunting and gathering societies offered women considerable voice, based on their important economic contributions to the group. In most agricultural settings, however, men claimed control over the basic crops, with women relegated to inferior status within the family and in society at large – thus creating the classic patriarchal system for the first time. Agricultural families also needed a higher birth rate, to generate labor for a largely family-based production system, and the functions of children were accordingly redefined.

And the list could easily be extended. Agricultural communities, for example, faced problems with contagious disease and recurrent warfare that had been less significant in hunting and gathering settings. On the other hand, in normal years they could generate some food surplus, which allowed a minority of people to specialize in occupations other than farming – from serving as priests to developing manufacturing skills as artisans. Small wonder, again, that most world historians emphasize the arrival of agriculture as a key turning point in human affairs.

But it is not clear that, where time is concerned, the level of change was quite so pronounced. Some connections certainly deserve exploration. Peasant villages, for example, were larger than the average hunting group: a few hundred people rather than a few dozen. This means that the easy personal contact among individuals in the group became more complicated, and more formal coordination might be necessary to assemble group members for shared tasks or ceremonies. Reliance on agriculture may have created even greater interest in precise calculations of seasonal change than had been true for hunters. Agriculture also encouraged new ceremonies – in the northern hemisphere, spring rituals to promote fertility of crops, fall celebrations of the harvest – that might have implications for calendar time. While early agricultural societies remained polytheistic, their beliefs often became more complex, their hopes to interpret the intentions of the gods more detailed.

On the other hand, many key time measurements were already available – most obviously, the ability to calculate and celebrate the major solstices and lunar cycles – and these could still be used and embellished. And while villages were larger and more complicated than hunting groups, they were not huge, and personal contacts could still render most formal scheduling unnecessary. The advent of agriculture, in sum, created some new needs and opportunities but probably not the kind of human revolution that occurred in other domains

For example, the Konso people in central Africa developed seven divisions of the day, between sunrise and sunset, ranging from one to three hours, and they had names for each; but their units were determined by simple observation and they were measured in terms of activities – one period for getting the cows to pasture, another for when they came back, long periods during the heat of the day when there was little activity. Near the equator, there was little seasonal change, so this case is a bit distinctive. But the general point, that agricultural people did not need a lot of innovation in timekeeping, largely stands.

But this is not the end of the story. Ultimately – another familiar world history marker – after some time most agricultural societies generated a further change, in creating more complex societies, or civilizations, marked by several key developments: the rise of more substantial and significant cities (even as the majority of the population remained rural); the establishment of formal governments to replace more informal leadership; and the creation of systems of writing.

These changes might have more implications for the sense of time than agriculture itself did. Cities, even though very small by our standards – possibly 30,000 people at an early extreme – might generate coordination requirements that depended on a wider shared awareness of time. Cities also fostered a larger number of skilled artisans, who might tinker with devices relevant to timing. Government institutions might create a need for clear scheduling. One historian of time, for example, refers to "bureaucratic" societies, and their more detailed interest in time, as a contrast to less structured groups. Most courts of law, for example, operated during certain times of the year, and also convened during particular times of the day. Urban governments sometimes established some rudimentary police force to prevent disorder after sunset – a night watch – which might need to be relieved after a set passage of time. Rulers might also claim control over a society's system of time as an exercise of power. Finally, writing – though long confined to a literate minority – created opportunities to record and transmit knowledge, with major implications for the study of different forms of time. While we will see that even in civilizations, many people continued to rely primarily on older notions of time – among other things, because 80% or more of the population remained rural – there is no question that a number of incentives and opportunities for change emerged.

One shift is interesting at the outset. While many early religions explored ideas about creation, wondering how the earth and its inhabitants began, many agricultural civilizations also developed an idea of some idyllic human past, which contrasted with the burdens of life at present. Judaism, and in its wake Christianity and Islam, explored the notion of an initial garden of Eden, from which humans later strayed. Chinese philosophy emphasized the wisdom of the past. Greek and Roman thinkers established the notion of an earlier Golden Age, which life was better and purer. Similar notions arose in a number of African religions. While a reference to a better past does not necessarily create a deep interest in the details of historical chronology, it does suggest some awareness of change over time and simply the existence of time in the past. More prosaically, many civilizations generated scholars eager to record developments from one point to the next – whether by year or by royal reign or by some other measure. Location in historical time might not matter much to many people, but some new consciousness emerged as societies became more complex.

Far more important, however, were the needs and opportunities agricultural civilizations created for additional kinds of time measurement. Two categories emerged, though they might interrelate. First, agricultural civilizations generated more sophisticated mathematics and astronomical observation, that could be applied to various kinds of time. And, second, they created new devices through which time, in its various aspects, could be assessed. The change here was real, as we will see: many of our current ideas about time trace back to achievements of early civilizations. At the same time, the rhythm of life, for most people, was not totally transformed. Many basic activities continued to be largely independent of any elaborate sense of time, beyond what was established by seasons or the daily position of the sun. Grasping the limitations of change, along with the key shifts, creates a challenging analytical framework for this period in human history.

Mathematics and Measurement: New Kinds of Precision

Agricultural civilizations had at least two reasons to develop new capacities in mathematics. They needed ways to calculate physical space, to determine the boundaries of property. And they almost always sought methods to observe and calculate the movement of stars and planets as these were believed to allow predictions of future events; a deep interest in astrology emerged in most early civilizations in Asia, Europe, and the Americas, and it would long persist.

Measurements of space were not directly related to time, but they might yield relevant categories; defining the properties of the circle, for example, could yield concepts relevant to annual or daily units as well; the applicability of units of 12 and 60, derived from Babylonian calculations of circles, showed up particularly in ideas about divisions of the day, the hour and, later, the minute. Divisions by 12 (or 2 x 12) and 60 turn out to work perfectly well, but they are of course human inventions, not edicts of nature. Babylonians also established a common measure of distance – called the Babylonian mile, though it was about seven miles long – that was later applied to calculations of the position of the sun. Astrology was even more directly connected to time, so that planetary alignments could be precisely calculated and astrological forecasts attached to calendars.[1] Both types of measurement, finally, promoted a more general interest in developing more precise mathematical categories that could create a greater sense of orderliness in human affairs and in nature alike.

Years and Months

Obviously, the agricultural civilizations had ample precedents in developing calendars, based on established knowledge of the lunar cycle or the solstices or both– and all of them worked intensely, though diversely, on establishing greater precision. All the early civilizations devoted considerable attention to clearer and more precise calendars. Royal governments were often directly involved as sponsors, both to assure the proper timing of major religious rituals and also, in many cases, to record the years of an empire or a reign with accuracy – a new factor, reflecting a novel interest in historical timing associated with political power. Sometimes, a new dynasty would deliberately replace an older calendar as a demonstration of leadership and a means of keeping track of the duration of the regime.

Specific results varied. With calendars, as with other measurements of time, key regions made initial decisions separately. Different geographies – for example, different relationships to the position of the sun; different specific agricultural systems and planting cycles; and, perhaps more important, different beliefs about cosmology, determined a wide range of different systems.

Thus, early Chinese calendars, before 770 BCE, were based on the annual movement around the sun and divided the year into ten months of 36 days each. Lunar divisions, yielding 12 or 13 months, were also developed, and ultimately lunar observation would play a vital role in relevant Chinese science. Calendars were of vital importance in India, as part of the emerging Hindu religion and the precise scheduling of sacred ceremonies. Scholars experimented both with lunar and solar measures, but on balance the lunar pattern was preferred. What resulted was a series of relatively short months, corrected every few years with an extra month to ensure that festivals and harvest celebrations occurred in the appropriate season. Buddhism would ultimately use the lunar approach as well.

Mayans, who devoted great attention to calendars and calculations of time both past and future, identified 18 months of 20 days each, with an additional five days annually to capture the full year. The extra five were associated with bad luck, and filled with prayer and mourning; anyone unfortunate enough to be born during this span was presumably "doomed to a miserable life". Mayan astronomers were also aware that a full solar year involved slightly more than 365 days. But the Mayans also had a different ritual calendar of 260 days per year, divided into 20 13-day weeks; days in each week were labeled for a god and used for prediction and schedule. Some days brought bad luck; some were all right but not suitable for marrying; and so on. This measurement was strongly attached to observations of the location of the planet Venus in relation to earth.

A society on the Indonesian island of Java (in the southern hemisphere) began its year with the winter solstice, but then had months that lasted between 23 and 43 days depending on changes in the sun's angle toward their particular location.

African traditions presented a variety of approaches, in addition to some really distinctive concepts of historical time. Seasonal factors were usually vital, often dividing a year into two parts, one with active agriculture, the other not; in some cases these basic units preempted any interest in the year itself. The Akun people, in present-day Ghana, in addition to emphasizing two basic seasons, developed a 40-day month, used strictly for political rituals. Interestingly, most African religions did not emphasize particular points during the day and, other than sunrise and sunset, daily divisions prompted little interest. Overall, African cultures suggested much less interest in time than was true of several other world regions, ultimately including the West.

Calendar development in Mesopotamia – ultimately, particularly influential – began on a lunar basis, again periodically adding an extra month to catch up to the cycle around the sun. No weeks were initially designated but holy days were usually scheduled on the first, seventh and 15th of each month. Later, however, particularly among the early Persians, emphasis shifted to the sun for annual timing – linked to intense worship of the sun. This yielded a 360-day year, with 12 months of 30 days each; but a 13th month was added every six years to synchronize the calendar with the seasons.

In sum, while specific patterns varied the early civilizations were all marked by careful calendar development and, usually, some effort to combine lunar and solar factors, guided particularly by the need for accurate ritual timing and, sometimes, the interest in defining the longevity of an empire. Weeks, particularly when they too were partially linked to lunar cycles, might factor in as well, as was obviously the case with the Jewish calendar. Some societies devised names for days within the week, often with religious meaning, but in other instances, as with initial Mesopotamian efforts, no names were applied; ritual days were simply counted off from the beginning of each month. The mixture of common interests and decisions and diverse specifics was intriguing, though there was a fairly widespread tendency to try to combine lunar measurements with solar modifications.

We do not know how many people paid much attention to this fascination with calendrical measurement in any of the societies involved, particularly given widespread illiteracy. Many might be content to be called to the major ceremonies by appropriate authorities or to go along with a local crowd, otherwise marking the passage of annual time by the observable signs of the seasons. But there is no question that, at some levels, the codification of calendars was a major achievement of the early civilization period.

Days and Hours

The interest in time measurement did not stop with calendars, however. Early civilizations also developed an interest in units of time during the day. The subject commanded less attention, partly because there was less ritual and practical need, partly because measurements had less to do with the movements of planets and stars. There was of course the daily rising and setting of the sun and, as we will see, the Egyptians used stars in assessing units of the night. As matters of simple observation, however, these daily passages might not inspire additional measurement. But a number of decisions emerged, particularly in Egypt and Mesopotamia, that still define many of our conceptions of time today.

The crucial question was how to chop up the day, beyond the distinction between night and light – a distinction that, to make things more complicated, varied with the season. As with calendars, different societies made different choices. Some Mediterranean societies, including Rome, divided the daylight period into hours but simply carved the period of darkness into thirds (roughly four hours each), to help schedule the night watch. Early Chinese calculations also split the night into watches of the guard (each covering a fifth of the time from sunset to sunrise), but marked the whole 24-hour span into a hundred parts. Southeast Asia emphasized fourths of the day, beginning at seven in the morning. A system in India identified 30 units in each day, of 48 minutes each (as we would calculate; the Indians did not initially define the minutes involved), divided further into 60 total units, each of 24 minutes. The interest in developing a clear system was obvious, but there were many ways to go about it. And it is at least amusing to remember that hours of the day, of whatever sort, is a human-made invention, even as most modern people organize much of their lives around them.

A number of early societies also shortened or lengthened their basic divisions according to the season. An hour in a winter day would thus be shorter than its summer counterpart, the key need being an accepted set of units for the daylight period. In many cases this reflected a deep attachment to the mathematical regularity of having 12 divisions for day, another 12 for night. It also raises real questions about how hours could be used in practice, if in fact their duration varied considerably.

For ultimately, it was the system developed in Egypt that proved most influential, initially in the Mediterranean basin and the Middle East, but ultimately more widely. Egyptians at first divided the daylight period into ten hours, but with another hour at each end to capture the transition to and from the presence of the sun. Later, this same 12-hour pattern was applied to the night as well, as the Egyptians came to calculate divisions here on the basis of the position of 12 stars – the overall result yielding the familiar total of 24 for the day as a whole. Why the Egyptians proved fascinated with units of 12 is a matter of some debate. Some scholars think that an interest in 12 initially followed from the ability to count to 12 on the joints of one hand (three joints in each finger). Other argue that the Egyptians simply applied to the day the same lunar unit – 12 – that worked fairly well for the year. In any event, 2 x 12 it was, first in Egypt, then in places like Persia and Rome.

And, characteristically, the Egyptians quickly connected their daily divisions to the gods: each hour was assigned to a particular divinity with appropriate prayers or propitiations available; and each hour was accordingly separately named. Again, the connection between early advances in concepts of time and religious needs proved vital.

Early civilizations did not go further than the hour, for greater precision was neither possible nor necessary. When – much later – hours were divided up, into units of 60 (each then further divisible by another 60), Mesopotamian mathematical principles were used, based obviously on units of 60 – the same calculus that had long been applied to measuring circles. But discussion of this nicety came well after the initial definition of units in the day – the first reference to "minutes" comes from Europe in the 13th century CE. It played no role in discussions of time conceptions through the early civilization period and well beyond.

For, quite apart from measurement challenges, most societies, even when accepting the notion of 24 hours in the day, long continued to separate day and night into equal parts. Since the duration of each changed with the seasons, the length of hours would vary markedly as well – long during daylight in the summer, in the northern hemisphere, but much shorter at night. The idea of dividing hours into standard, precise units was a non-starter in this context. At most, measurement devices might allow divisions into halves and quarters (again, of variable durations). All of this, for better or worse, was a far cry from the kinds of precision that would emerge in more modern times – leaving people in the early civilizations less burdened by the details of daily time, but less able to measure accurately as well.

Measurement and Devices: More Innovation

Early civilizations developed three major ways not just to talk about time during the day, but to chart it directly. The most basic approach highlighted the sundial, which could be readily linked to the discussions about hours during daylight. The challenge of time at night was ultimately met, particularly by water clocks, long a vital kind of device. Finally, a number of other instruments were introduced to measure duration within a day or night, with no necessary connection to more elaborate systems of timekeeping.

Sundials were devised in many early civilizations and, as noted, certain crude versions may even have been used by some hunting and gathering groups. There is some debate about details. There is evidence of sundial use in Egypt as early as 3500 BCE, though the more refined devices emerged about 2,000 years later. Egyptian obelisks, the tall, angled towers located near many temples, may have been used as sundials, but the balance of opinion now inclines against this. The Greek historian Herodotus thought that sundials were a Mesopotamian invention, claiming that Greeks learned of them through this contact (but not until about 560 BCE) – and certainly Mesopotamian use was active. Mesopotamians in fact used a distinctive term: shadow clocks. Early Chinese societies introduced sundials, but there is not a lot of information about dates and precise techniques. An early sundial has also been found in what is now Russian territory.

Sundials are at once a very simple and a complex device. Simple in that the basic idea is simply to put up some marker that casts a shadow from the sun, and then construct a base, called a dial, with markers that tell what time it is when the shadow hits it. In a Greek play, for example, a wife asks her husband to come home when the shadow hits ten feet. But the complexities involve more than measuring shadows. In the first place, sundials long had to be explicitly calibrated for a particular spot, since the angle of the sun varies with latitude; this called for skilled craftsmanship. Further, during the day itself, the marker, or gnomon, has to be moved at least once, from morning to afternoon, again to pick up the sun's shifting position; Egyptians presumably employed people to do this regularly, in the dials located in public places. Finally, dials themselves have to be carefully calibrated; this is where the notion of specific hours of the day would be very useful, as in Egypt, and further notes about half or quarter hours could be introduced as well.

In modern times, it's fairly easy to buy a simple sundial for home decoration. But it is unlikely that these complicated devices were widely available in the early societies, though Egyptians did make some portable models. The instruments were normally located in major cities, in prominent public spots with an elaborate, marked dial often formed by crafted tiles or polished stones. Sundials figure in the Old Testament of the Bible, in the book of Isaiah, when God promises to turn back time to help a king: "Behold, I will bring again the shadow of the degrees, which has gone down in the sundial of Ahaz" – a Hebrew king who reigned around 700 BCE.

Sundials obviously had a huge limitation: they were no good at night. But many early civilizations wanted some calibration for the night as well, either to feed mathematical calculations about time or for very practical purposes, like making sure priests were ready to conduct a religious service at the crack of dawn or night watchmen relieved at set intervals. To be sure, Egyptians ultimately developed modifications of the sundial to measure positions of the stars as well, drawing "nighttime" lines on the dial; but it was hard to be accurate. The more common response to the challenge of measuring time after dark was the water clock or, in Greek, clepsydra.

Water clocks may also first have been invented in Egypt. The oldest clearly documented clock dates from about 1400 BCE, where it was used in the Temple of Amen-Re at Karmak. Egyptian priests became skilled technicians, in order to assure that temple rites and sacrifices could be performed at the correct time, even at night. Water clocks were also widely introduced in Mesopotamia; a document from the 7th century BCE refers to the use of these clocks to determine the pay owed to guards during both night and day. But China may have a lead in this category – one claim goes back to 4000 BCE, and certainly the Chinese, through premodern history, would invest greatly in this device. Early water clocks were also developed in India: a document from the second millennium BCE refers to a clock that measured time in intervals of 24 minutes. Water clocks in Persia, a bit later, were used to determine the timing of religious ceremonies; they were also widely deployed to regulate the amount of water a farmer could take from an irrigation system – the individual would have a set period of time and no more.

The basic device was simple enough. A primitive water clock may involve simply putting a set amount of liquid in a pot, letting it flow out gradually and evenly, and measuring the result. The clocks can initially be calibrated, during the day, by use of a sundial – allowing experts to determine how much water would flow in a particular unit of time. Clocks can also be constructed by measuring how much water flows into a container, with lines marked to tell how much time has passed. In many early water clocks, separate columns were used in the container, each marked for the season – to take into account the fact that the length of hours varied seasonally.

Water clocks could be used simply as timers, rather than devices to chart divisions during a whole day; this may in fact have been the most common application. Here, owners could merely decide what interval they wanted to allot to some activity, whether they had a concept of hours or not, and then fill the device accordingly. This approach probably applied to the use of water clocks to regulate the duration of an individual's access to an irrigation system. But wider applications developed. In classical Athens, some brothels used water clocks to make sure no client overstayed his welcome.

Like sundials, most water clocks were deployed in public buildings, for they required a great deal of care for accuracy – though problems like changes in the rate of flow when water levels dropped were hard to compensate for entirely. Often, large bronze pots were used, and great skill went into providing accurate markings. While the main use of these clocks was at night, they could be deployed during the day as well – for example, to compensate for a cloudy sky when sundials were useless.

A final set of early devices emerged not to tell time in a daily sense, but to determine set intervals by which the duration of an activity could be measured – rather like a very early version of a stopwatch. A classic device, once glassblowing was introduced by the 14th century BCE (in Egypt again), was the sand glass – what today we call an hour glass. Sand glasses involved a tapered cylinder that permitted a slow, gradual flow of sand into another cylinder – when the first part was emptied, time was up. Still other devices to determine intervals included oil lamps – where a particular amount of oil would be measured out and then burned until it was used up; or simply a marked candle (though this could in fact burn unevenly).

Clearly, a great deal of ingenuity was deployed, in all the early civilizations, to improve on natural signs of time, particularly during the day,. Many of the technologies linked to the efforts to define units mathematically, translating theory into daily practice. But, obviously, many of the results were somewhat imprecise – even determining a quarter of an hour required elaborate calibrations on a sundial or water clock. And the best devices assumed considerable investment and technical expertise. The desire for some regularity was clear, but the impact, particularly outside great temple cities, is harder to assess. The lack of many devices, except perhaps for very simple sundials, for household-level use is obviously fundamental.

Uses of Time in Early Civilizations

What kinds of time affected ordinary people as the agricultural civilizations took shape? Did the new calculations and devices have much impact? The questions are important, but hard to answer, simply because of lack of much direct evidence.

There is certainly every reason to be cautious about projecting much basic change. Some of the most interesting innovations may have had had little to do with rigorous measurements, though they might reinforce the natural markers already available.

For the vast majority of people, in the civilization zones, still lived in rural villages, distant from the calculations and the timing devices. They worked more hours a day, on average, than their hunting and gathering ancestors had, but the work was variable and highly dependent on natural cycles. Most obviously, little work could be done when there was no daylight. Even in peak agricultural seasons, during planting and particularly harvesting, people would depend on sunrise to get going, and would call it quits when darkness settled in.

Furthermore, peak periods were unusual. During much of the year, and particularly of course the long months between harvesting and new planting, work days were short and often irregular. There was no timed schedule involved. Furthermore, most peasant labor involved little coordination beyond the family unit, though harvesting might be done in larger groups. And while villages, with 200 to 400 people, were larger than the hunting groups had been, informal contacts would be sufficient for most collective activities; again, timing devices or signals were not essential.

Many agricultural villages did experience one change that might reinforce the natural rhythms of day and night: they acquired some domesticated animals, many of which had daily cycles of their own that might affect the people around them – particularly when they involved noise. World history rarely includes much attention to our feathered friends, but attention to time encourages an exception.

For ultimately, in most rural regions, roosters would gain special importance for their daily wakeup calls. There is evidence that some kind of chicken was domesticated in southern China as early as 6000 BCE. But, more probably, chickens were first widely used in India, possibly around 4000 BCE. Initially, it was cockfighting, not food, that drew the greatest interest: but, on this base, domesticated chickens began to spread fairly rapidly, from south Asia to other parts of the continent, even before the apparatus of civilization developed in some of the regions involved. Chickens had spread to southern Europe by 3000; Phoenician traders carried knowledge of the fowl widely around the Mediterranean, though the bird became known to Western Europe only after about 1000 BCE. The arrival of chickens in sub-Saharan Africa is less clear, though it did occur; and there is debate as to whether some knowledge of at least a type of chicken may have spread from Polynesia to South America before the arrival of Europeans. But the main point is clear: chickens became widely popular, ultimately primarily as a food source, becoming standard features of most rural villages and many cities as well.

And with roosters came the predictable salute to dawn. Roosters have an innate rhythm that pushes them to crow as the day begins, presumably to help stake out territory and warn off rivals. Daily contact with this familiar prompt became a standard part of agricultural life – and possibly the most significant difference, in terms of daily routine, from the experience of hunters-gatherers: it may have become a bit more difficult to sleep in, even on cloudy days. There was scant need for any other timed reminder. As we will see, some groups actually carried roosters with them when they traveled, as a wakeup device. But obviously, roosters did not promote any particular timed divisions of the day after sleep had ended.

Popular leisure, also, reflected the lack of any particular commitment to rigorous timing. Two kinds of distractions from work were common in peasant societies. First, during the work day, many people would take some irregular time off, to chat or wander around or, often, take a nap. Again, there were periods when this was impossible, except for real slackers; the work demands were too great. But what many modern people would view as a surprising lack of rigor or discipline was part of the normal experience for many people, rural and urban alike; and there was no scheduling involved.

The most elegant and cherished leisure form, in agricultural societies, was the communal festival, and depending on resources and local culture there were usually a number of festival days each year. Many festivals, of course, featured religious rituals. Some were associated with the agricultural cycle, around planting and harvesting. Others recognized important military victories – in Egypt, for example, the pharaoh decreed popular celebration to commemorate the defeat of a Libyan army. This means, of course, that calendar time was often involved; but clock time was not.

Whatever the initial justification, most festivals involved a variety of communal activities beyond any specific rituals. Usually they offered dancing, group feasting, games and contests – like wrestling matches or tugs of war between bachelors and young husbands. Again, the specific roster depended on the local culture. These events spread through the day, with no set schedule, and in some cases into the night, illuminated by bonfires. Since most activities came from the group itself, rather than trained performers, they could take shape somewhat spontaneously once the celebration began. Even eating often spread through the day, depending on impulse.

Leisure, in other words, largely confirmed the absence of any rigorous timing from village life. To be sure, many festivals themselves depended on some awareness of annual calendar, though some of the biggest celebrations, around the winter solstice for example, continued to depend on basic calculations of natural cycles. Many villages had a least one or two literate members, often religious officials, who might help with this kind of calendar-keeping. And, as already suggested, most adult villagers were surely aware of the week, if only to specify a particular day for market activities, whether the pattern had spilled over from hunters-gatherers or was devised as part of the agricultural routine.

None of this should paint an unduly idyllic picture, despite the probable absence of huge time constraints. Village life could be hard, with demanding physical labor, even if explicit time pressures were usually absent. Many early civilizations also kept slaves and, while their work was often varied, some were subjected to harsh discipline. Meager diets and high rates of child mortality attest to the rigors of agricultural life. But on the specific subject of time, there was often some latitude. To be sure, there were situations where some sense of time became more important. Calendar decisions about when to plant crops might link peasants with more elaborate timing schemes; and where vital resources, like irrigated water, were doled out by time slots, as in ancient Persia, even more precise timing might be crucial. Still, most villages long lacked any significant timing devices, and most villagers probably felt no need for them.

Conditions were somewhat different in the cities – though not entirely distinct. Cities, of course, were where governments or religious authorities made their big investments on timing devices. But even in the cities most work took place in a family context. Artisans, for example, worked in a shop attached to their home. The work group consisted of other family members plus, often, a few outsiders – including apprentices but also another laborer or two. But these people, as well, usually lived in the family quarters. Getting the work group going, in other words, required no elaborate coordination. Clock time was rarely required: sunrise and perhaps a neighborhood rooster would usually do the trick. By the same token, craftsmen might open their shops to customers without any sense of rigid schedule. In many cases the lack of a standard definition of the length of an hour might complicate practical timing as well. Weeks mattered in the urban routine, to signal market days, but the day itself was less rigidly defined. For many ordinary people, even prayers and religious ceremonies were conducted largely within the family, not depending on wider coordination.

However, as civilizations developed, cities did generate some new needs for a clearer sense of time. This is why many cities might feature public sundials of some sort, and why oil lamps, sand glasses and water clocks gained attention.

Two or three specific needs emerged. The most obvious, at least in some societies, was religious. Some civilizations, at least, generated beliefs that certain rituals, honoring the gods, had to be performed at particular times of the day. This was most apparent in ancient Egypt. Priests and, sometimes, the pharaoh himself would participate in services at daybreak. In temples across Egypt, a high-ranking priest would wash and elaborately dress the god's statue and present it with gifts – presumably, to help assure a successful day ahead. Obviously, it was important to time the ceremony appropriately, with officials ready before dawn – and this meant, in turn, that some measurement of time had to occur during the night, appropriately adjusted to the changing seasonal patterns of sunrise. Here, in sum, was a very practical need for devices like the water clock.

In Egypt, at least, these daily events were confined to a small group of priests, again sometimes including political leadership as well. They were not open to a wider public, which was thus free to follow the somewhat less rigorous patterns of the workday. Seasonal celebrations were another matter, where days were designated for larger assemblies to pray and offer ritual gifts to the gods. And while some of these days might link to natural cycles, like the solstices, they depended more on the annual calendars that all the major civilizations developed.

A second need for more precise timing involved the desirability, in some circumstances, of designating precise starting and stopping points for a particular activity. This was the source of interest in devices such as sand glasses, for example, or oil lamps. There are a number of examples, though it is not always clear when, chronologically, these needs were identified in the early civilizations. Thus is might prove desirable to time a public speech or a presentation in a court of law, to make sure the official did not drone on endlessly. (Timed speeches would become part of the operation of democratic assemblies in Athens, for instance, at a somewhat later date.) Or particular kinds of torture might be timed, so that they did not seem too cruel or arbitrary. Or, in cities that organized nightly patrols for safety, it might be important to measure intervals so that the guards could be relieved before they succumbed to sleep. Later on for example, as in colonial New York, members of a night watch actually carried a small sand glass, so they would be able to determine when their shift was over. Again, we lack detailed records of this kind of development in the early civilizations, perhaps because some timing arrangements seemed so obviously practical to the authorities involved that no particular comment was required. And note that this kind of timing did not require division of a day into set intervals, for start and stop times were the only focus. But this kind of need clearly explains the reason that interest in timing devices often overshadowed efforts to chart the passage of time through a full 24-hour cycle.

Finally, though even more tentatively, some of the institutions of early civilizations might have generated an interest in arranging agreed-upon opening times, as determined perhaps by the various urban sundials available. Courts of law, for example, which involved gathering a number of different participants, often from various parts of the city, might have generated this kind of scheduling need. The same need might apply to some of the larger religious ceremonies: when would people be expected to show up during the appropriate calendar day? For the most part, however, we simply do not know when this kind of precision emerged – it would be clearer in more advanced civilizations later on, as we will see in the following chapter. Thus both Roman and Aztec societies stationed callers, in the urban centers, to note the time at which markets or other institutions would open. But in earlier cities, many group events may well have had rather casual starting points, with people simply following the flow of a crowd. A number of group activities, like law courts or royal audiences, operated with very little precision (kings, after all, could hardly demean themselves by bowing to clock time, even when it became available). Whether there was a start time or not, most of the people involved had to expect to wait their turn with no assurance of timing – a feature that still bedevils much public life today (as in most courts of law), in defiance of the modern preoccupation with clock time. Patience may have been easier to come by in societies that did not insist on timing in other respects.

Time has a rich history during the long early stages of the human experience, though there is much that remains unknown or imprecise. Interest in time follows from observations of nature and certain basic human needs. Calendar time long generated the greatest attention, both because of practical interest in anticipating seasonal change and because of the ritual needs of many early religions. But other kinds of time generated interest as well, particularly once more complex societies began to emerge.

Overall, a deep-seated human interest in measuring time and defining units balances against the fact that, well into the history of civilizations, daily needs for time, on the part of most people, might actually be rather modest – beyond what was naturally provided by seasons and sunrises. Thus the interest in calculating time outraced changes in the ways most people actually used time. And the calculations themselves were sometimes surprisingly vague – as in the persistent imprecision in defining the lengths of hours in the day. On the other hand, some of the most practical innovations, like the week, did not always generate a wider theoretical framework. The fact that interest in devices that determined starting and stopping intervals preceded more than crude attempts to divide the whole day into units of time is another indication of the very different patterns of life that prevailed among our ancestors.

The early evolution of time highlights one other important feature, though not a surprising one. Different regions and cultures, working independently, generated highly varied specifics when it came to time. Calendars differed, weeks might diverge, basic calculation frameworks reflected very different starting points. Here was a recipe for no small amount of confusion when societies came into greater mutual contact – and elements of this confusion linger even today, as an interesting complexity amid globalization (see Chapters 6 and 7, below).

On the other hand, reflecting both a common human nature and common human needs, plus the shared natural environment, there were some vital overlaps among the regions. The fact that a week developed almost everywhere is more important than its varying duration. The role of religion and ritual needs in determining many developments in the human approach to time is striking, again cutting across cultural specifics. The gap between theoretical calculations of time and the role of time in the daily routines of most people is also a common thread among the various early civilizations. These shared components help explain, as well, why information about time, and timing devices, was often eagerly shared when different peoples did come into contact. This aspect, the role of contact in shaping approaches to time, would become even clearer in the next major phase of the human experience. The fact that so many societies developed human constructs to deal with time, beyond their efforts to calculate rhythms of nature, was another common feature that could, at various historical points, prompt mutual interests.

Note

1 On the complex relationship of astrology and astronomy, see http://en.wikipedia.org/wiki/Astrology.

Further Reading

Adjaye, Joseph K., ed. *Time in the Black Experience.* (Westport, CT: Greenwood, 1994).

Aveni, Anthony. *Empires of Time: Calendars, Clocks, and Cultures,* revised edn. (Boulder: University Press of Colorado, 2002).

Aveni, Anthony. *Skywatchers: A Revised and Updated Version of Skywatchers of Ancient Mexico.* (Austin: University of Texas Press, 2001).

Bourdieu, Pierre. "The Attitude of the Algerian Peasant," in J. Pitt-Rivers, ed., *Mediterranean Countrymen.* (The Hague: Mouton, 1963).

Bruton, Eric. *The History of Clocks & Watches.* (Edison, NJ: Chartwell Books, 2004).

Colson, F. H. *The Week.* (Cambridge: Cambridge University Press, 2015).

Cotterell, Brian, and Johan Kamminga. *Mechanics of Pre-Industrial Technology: An Introduction to the Mechanics of Ancient and Traditional Material Culture.* (Cambridge: Cambridge University Press, 1992).

Cowan, Harrison J. *Time and Its Measurement.* (Cleveland and New York: The World Publishing Company, 1958).

Evans-Pritchard, E. E. *The Nuer: A Description of the Modes of Livelihood and Political Institutions of A Nilotic People.* (Oxford: Oxford University Press, 1940).

Fermor, John. "Timing the Sun in Egypt and Mesopotamia." *Vistas in Astronomy* 41, no. 1 (January 1, 1997): 157–67. https://doi.org/10.1016/S0083-6656(96)00069-4.

Fermor, John and John Steele. "The Design of Babylonian Waterclocks: Astronomical and Experimental Evidence Centaurus." *International Journal of the History of Mathematics, Science and Technology* 42 (2000): 210–22.

Hawkins, Gerald S., and John B. White. *Stonehenge Decoded.* 2nd edition. (New York: Hippocrene Books, 1988).

Kumar, Narendra. *Science in Ancient India.* (New Delhi: Anmol Publications Pvt Ltd, 2004).

Landels, John G. "Water-Clocks and Time Measurement in Classical Antiquity." *Endeavour* 3, no. 1 (January 1, 1979): 32–37. https://doi.org/10.1016/0160-9327(79)90007-3.

Marshack, Alexander. *The Roots of Civilization: The Cognitive Beginnings of Man's First Art, Symbol and Notation.* Revised, Expanded, Subsequent edition. (Mount Kisco, NY: Moyer Bell Ltd, 1991).

Needham, Joseph, Wang Ling, and Derek J. de Solla Price. *Heavenly Clockwork: The Great Astronomical Clocks of Medieval China*, 2nd edn. (Cambridge: Cambridge University Press, 1986).

Needham, Joseph. *Science and Civilization in China: Volume 4, Physics and Physical Technology, Part 2, Mechanical Engineering.* (Cambridge: Cambridge University Press, 1965).

Parker, Richard A. *The Calendars of Ancient Egypt.* Studies in Ancient Oriental Civilization, no. 26 (Chicago, IL: University of Chicago Press, 1950).

Richards, E. G. *Mapping Time: The Calendar and Its History.* Revised ed. edition. (Oxford: Oxford University Press, 2000).

Whitrow, G. J. *Time in History: Views of Time from Prehistory to the Present Day.* (New York: Barnes & Noble, 2004).

A number of psychologists and linguists have also been studying the presence or absence of any sense of historical time among existing hunting and gathering cultures; one cultural cluster in Brazil for example has memories but no belief in time as a continuum. See Vera da Silva Sintra, "Event-Based Time in Three Indigenous Amazonian and Xinguan Cultures and Languages," *Frontiers in Psychology* (2019).

3 Time Amid the Classical Civilizations and World Religions

A Roman writer, Aulus Gellius, generated a surprisingly modern-sounding time lament, around 184 BCE, claiming that he much preferred his childhood when his sense of time was regulated, quite adequately, by the grumblings of his stomach: "The gods confound the man who first found out how to distinguish hours – confound him, too, who in this place set up a sundial. To cut and hack my days so wretchedly into small pieces."

This chapter covers both the age of the great classical civilizations – Persia, China, India, Greece/Rome – but also the subsequent world history period marked by the spread of Buddhism and Christianity and the rise and rapid expansion of Islam. The classical societies began taking shape around 800 BCE and would variously flourish for the ensuing millennium. Then, after the eclipse of the major classical empires by 500 CE, world history increasingly organized around the flowering of the great missionary religions, until the 15th century. A span of over 2,000 years constitutes a big chronological chunk, in terms of the standard framework for world history overall, but it makes sense for the key developments concerning time – after the innovations of the early civilizations. A number of important changes can be assessed, but most of them took shape gradually and, with a few important exceptions, their impact was not revolutionary. While some urban Romans clearly complained about a new sense of time, many people as late as 1400, still peasants, were not affected by time much differently from their ancestors, except perhaps in their religious lives.

As the classical civilizations matured, they continued to work on the calendars and time divisions inherited from the previous period, making important refinements particularly on the calendar side. Early Christianity, fusing with Roman patterns, sponsored some calendar decisions that are still in effect today – now, on a worldwide basis. Islam also introduced durable categories. Great attention was also devoted to water clocks and other devices, with some interesting improvements – though not fundamental new departures – to be noted. Growing contacts among the major civilizations, and particularly interactions between China and the Middle East and the Middle East and Europe, allowed some sharing both of concepts and technologies, again suggesting the continued eagerness to refine the approach to time. A number of societies expanded the production, size and use of bells, which (particularly in the cities) helped connect time to aspects of daily life.

The most striking innovations concerning time continued to come from the religious domain. Both Christianity and Islam introduced new emphasis on time during the day. The result was a new time sense at least among part of the population, as well as new encouragement to improve measurement and signaling. Continued interest in astrology, though less novel, also promoted attention to time and its relation to astronomy. Buddhism, the other great missionary religion, was less time-focused, but it too had some time concerns that could spread beyond purely religious rituals.

The role of the state also warrants attention, sometimes in tandem with religion. Even in the early civilization period, governments often flexed their muscle by imposing decisions about time as a means of illustrating their authority and organizing part of the population. During the classical period, and even beyond, particularly in empires like China or the Incas and Aztecs, this process continued.

The period under consideration also saw the development of important new contacts among major parts of Africa, Asia, and Europe: the famous Silk Roads introduced new overland exchanges, while later a whole series of routes linked to sea lanes in the Indian Ocean, and trade and other interactions intensified. Here was the basis for some of the sharing of timing mechanisms. It also created both need and opportunity to coordinate some aspects of time measurements through larger regions – particularly when it came to calendars. But the exchanges were not yet extensive enough, and above all not rapid enough, to generate any sense of a need for detailed time coordination across regional boundaries. This issue would emerge only in the next period, after the 15th century, and became really pressing only in the 19th century, with the transportation and communication technologies birthed by the industrial revolution. Travel did expand particularly during the millennium before 1500, but mainly by ship or pack animal, it did not really challenge separate regional patterns of timekeeping.

The timespan running roughly from 800 BCE to 1400 CE was still pre-clock, in terms of the technologies available to most people. The period saw important new thinking and decisions applied to some aspects of time, within the framework of available calendars, sundials, and water clocks. The importance of coordinated timing in the cities measurably increased, as the Roman lament suggests. For some, a new religious fervor further redefined the daily need for time, providing the most striking innovations in human scheduling.

Calendars

The great classical civilizations and their successors explicitly built on many markers that had already been established. Of the various categories of time, calendars were in many ways the most advanced, requiring the least additional work. Nevertheless, a number of new decisions were introduced, often under government sponsorship but increasingly reflecting the impact of the major religions as well.

In many different societies, ongoing attention to calendars reflected a scientific thirst for precision, often linked to the desire to anticipate religious ritual events with the greatest possible accuracy. In some cases it can also be suggested that laying out calendars represented not only a way to anticipate future schedules, but to claim some control over the future as well, some means of reducing unpredictability by mapping the months and years to come.

Long past was the time when great effort was required to identify the seasonal solstices, as in erecting massive monuments; some earlier constructions now passed into disuse, or were even raided for building stones. Available astronomical knowledge pinpointed the key dates precisely, and also charted other potential markers such as eclipses. Solstices still might be the subject of celebration –the winter solstice remained particularly important, for example in the great Roman festival, Saturnalia, and the later Christian decision to schedule Christmas at roughly the same point. But the calendrical calculations, themselves, were now almost routine.

Challenges remained, however, particularly as scientists sought greater precision. Existing adjustments of a lunar calendar, by adding a few days either every year or every few years, still did not capture the earth's cycle around the sun with complete accuracy. (The adjective for introducing day(s) or months(s) periodically into a standard calendar to harmonize with the solar year is *intercalary* – as in February 29 in the Gregorian calendar today.) Another pressing issue in a number of regions was the diversity of calendars in use, from one locality to the next. As political empires or religious systems expanded, efforts at standardization were almost inevitable. Finally, because calendars served diverse purposes, some early societies had developed different systems within the same region. Thus early Greeks had one calendar, based on lunar observations but modified by adding a few days a year, to regulate ritual ceremonies; but another, much shorter, was used to calculate the dates of annual political events. This kind of confusion needed sorting out.

There is no need to go through all the various regional improvements, but a number of adjustments represent typical types of reform, some with lasting consequences. Striking variety and considerable experimentation persisted, around the basic commitment to astronomical patterns. A maverick Icelandic calendar in the 10th century argued for 52 weeks or 364 days, based on purely solar reckonings; periodically a "leap week" would be introduced for adjustment purposes.

Asia

Persia had a long history of calendar calculations, with periodic adaptations depending on religious and political factors. A new dynasty in the 11th century commissioned al Khayyam, a leading astronomer, to improve on the data on which the calendar was based. Khayyam was joined by other astronomers at what was at that point a state-of-the-art observatory. The team worked for 18 years, advancing the accuracy of existing astronomical tables and developing a solar-based calendar of unprecedented accuracy, still officially used in Persia and Afghanistan. According to the new calculations, the actual length of a year is 365.24219656156 days. This calendar, called the Solar Hijri, features six months of 31 days each, followed by five of 30 days, and a final month of 29 (expanded to 30 in a leap year). The result is literally the most accurate in the world: in contrast, the Gregorian calendar widely used in the West and elsewhere is off by a day every 3,236 years – or, in every year, by 27 seconds.

Both India and China continued to sponsor calendar research. China's calendar still adhered closely to the lunar cycle, compensating by alternating 12 and 13 intercalary months to adjust to the actual length of the solar year. The process of adding a short month was initially somewhat haphazard, but became increasingly formalized. Thus the Han dynasty regularized the computation, dividing the year into 24 units, each paired for the resultant 12 months. A shorter additional month made up the difference from the lunar cycles. This traditional calendar continues to be used for the timing of festivals, such as the Lantern Festival, and also the designation of each year by a sign of the zodiac. As China's cultural influence expanded in East Asia, Korea, Vietnam, Japan, and Mongolia also adopted some version of the Chinese calendar, a crucial example of the role of new contacts in shaping the uses of time.

Research continued after the revival of imperial institutions in the 6th century CE. It has been calculated that over 100 different calendars were produced in China between the 4th century BCE and the 18th century CE. Emperors competed to sponsor more precise calculations of months, years, and also eclipse intervals, each hoping to outdo his predecessor. Chinese astronomers realized that the lengths of a year could vary slightly, several centuries before the same discovery in Europe.

Preoccupation with lunar cycles persisted: "The moon in her waxing and waning is never at fault…./Here at least is a trustworthy mirror on earth/To show in the skies never-rushing and never delaying." But in their fascination with imposing order – one historian sees the Chinese as trying to bureaucratize calendar times, as they did in their political institutions – scholars and officials also developed other divisions. For some calculations they divided the year into 24 15-day segments, each segment labeled according to its best purpose: thus there were particular segments in which it was optimal to plant, others to harvest, and so on. They also continued to utilize a ten-day week for some purposes. Finally, like the Mayans, they also tried to project time segments into the future, particularly emphasizing what they saw as recurring 60-year cycles in which key patterns might be repeated.

Buddhist calendars, building on Hindu patterns, also emphasized a solar-modified lunar pattern. Establishing certain dates with precision was religiously important, for example in designating the first full moon in the month of May as a time for particular commemoration of the Buddha himself. The Buddhist lunar calendar was quite complicated, with the periodic addition of days or even an extra month over a 57-year cycle. Different regional variants of Buddhism offered different dates for the beginning of the year, though a similar pattern of calculation operated otherwise.

Southern Europe

While developments in Asia were revealing, ultimately the most important, or at least influential, refinements of calendar calculations occurred in Greece and Rome, with ensuing modifications in European Christianity, and in Islam. Classical Greece continued to use a lunar approach, with an additional, or intercalary, short month for adjustment purposes to match the solar year. Accurate designations of the four-year cycle that set the time for the Olympic games provided an example of how calendars were becoming more important for coordination. But the Greek festival schedule was not based rigorously on the regular calendar, and there was also a political calendar that featured months of varying lengths. The overall result, by modern standards, was quite confusing; and each city state tended to have its own specific versions.

The Roman period, though using Greek precedent, ultimately generated a more orderly system. A traditional Roman year had ten months, with 304 days, but an early king added two months (Januarius and Februarius) to bring the pattern in line with the solar cycle. This evolved into a more regular 12 lunar months with an intercalary month of 27 or 28 days inserted at the end of February every three years or so.

However, the intercalations did not occur systematically, but on order of the governing ruler. And this created three problems. First, the system was so irregular that it was often well off schedule; on occasion, two intercalary years – that is, two successive years with extra days – had to be decreed simply to catch up with the seasonal flow. Second, individual governors sometimes prolonged the year simply to stay in power longer (or maintained the shorter year to punish a rival). And, finally, the decision on a longer year was frequently made without much notice, which meant that the average citizen frequently could not predict dates in advance. All these problems intensified in the period after 63 BCE, until Julius Caesar assumed power.

And it was Caesar who finally firmed up and clarified the arrangement, during the third year of his consulship. The year was to begin January 1, and run for 365 days; Caesar's effective successor, Augustus, added the notion of a leap year day every four years (maintaining February as the adjustment month), again to provide greater accuracy. This new, Julian calendar is still used in some branches of Christianity, and was widely deployed throughout Europe for many centuries. It was also under the Romans that the seven-day week, devised in the Middle East, gained wider currency.

Early Christian leaders introduced a variety of calendars. Their most important innovation, by the 6th century CE, was to establish a historical separation between the time prior to the birth of Christ (Before Christ, or BC) and the time "in the year of our Lord", or Anno Domini, thus positing two major eras in the human experience. This pattern caught on only gradually – historical time continued to be recorded by the reigns of kings or popes. (Spain for example long calculated from a beginning point of the "era of the Caesars".) But between the 11th and 14th centuries the BC/AD divide became fairly universal in Western Christendom. Christians also worked to assign a date to the beginning of creation: the Eastern Orthodox Church seized on 5509 years BC, the Coptic Church 5500, and, later, the Church of England postulated 4004.

The Catholic Church made one final move, in 1582, under the leadership of Pope Gregory XII, mainly to regularize the scheduling of Easter and to utilize improved astronomical calculations in Europe. To take into account the fact that the average solar year is just a bit shorter than 365.25 days (which had long been known), the Julian calendar was modified to eliminate leap days three times every 400 years, at the turns of the century. And, on a one-time basis, October 15, 1582, followed October 4, to compensate for the drift that had occurred since the advent of the Julian calendar. This new, Gregorian calendar was gradually accepted throughout most of Christianity (first, the Catholic countries; then, more grudgingly, the Protestant) and, ultimately virtually the entire world (Greece was the last European country to adopt it, in 1923.)[1] Overall, the history of Roman and then Christian calendars shows the importance of political and religious leadership in regularizing calendar patterns, and the utility of having agreed-upon calendars, in turn, for wider coordination within large regions.

Even before the Gregorian modifications, thanks to the Julian framework and Christian adoption, Western Europe may have benefited from the fact that a calendar year, with its definite beginning point each January, had been firmly identified. The growth of merchant activity in Europe by the 11th and 12th centuries, which included increasingly careful accounting procedures, depended on a generally accepted annual schedule, and agreed-upon beginnings and endings of the year. European merchant interests would play a more obvious role in decisions about daily time, as we will see, but a favorable connection with calendar time may have developed as well.

Islam

Islamic approaches to calendar issues took a very different path from the Christian options, but were also widely influential. Before the Prophet Muhammad, Arabs had developed a number of calendars, usually combining lunar months with a periodic intercalary adjustment to match the solar year. Great emphasis was placed on labeling each lunar month and identifying appropriate activities: most notably, war was forbidden during several months (initially, perhaps, to protect the food-growing season), and another period was designated for pilgrimage. One tribe was particularly responsible for designating the forbidden months.

The most obvious change introduced by the advent of Islam was a new dating system – the Hijri calendar, similar in broad concept to Christian calculations from the presumed birth of Christ. Human chronology was now divided between before and after the Hijra, or Muhammad's migration from Mecca to Medina. Dates are thus commonly listed as H, or in European languages AH (After Hijra), or BH (before). Year zero was, in Western calendar terms, 622 CE. By later 2019, the corresponding Islamic year was 1441.

But in most ways the more striking innovation, still powerful today, involved the belief that God disapproved of intercalary units, identifying exclusively with a lunar approach. As an early writer put it:

> The number of the months, with God, is twelve....Know that intercalation (*nasi*, in Arabic), is an addition to disbelief. Those who disbelieve are led to error thereby, making it lawful in one year and forbidden in another in order to adjust the number of the months made sacred by God....The evil of their course appears pleasing to them. But God gives no guidance to those who disbelieve.

Muhammad himself, in his farewell sermon, emphasized that "Nasi is an impious addition, which has led the infidels into error....The infidels profane that which God has declared to be inviolable." The Prophet argued that, in insisting on 12 lunar months, unadjusted, time was returning "to such as it was at the creation of the heavens and the earth".

Reliance on an unmodified lunar calendar (each month with 29 or 30 days, depending on the lunar cycle) had three results. First, the Muslim year was several days shorter than the solar year, and this remains true today. Hijri calendars thus cannot be easily compared to their Western counterparts, quite apart from the fact that they use a different base date.

Second, the older Arab tradition of identifying some months as particularly sacred or forbidden was maintained in principle. While Muhammad granted the validity of fighting off infidel attack during these months, they should otherwise be kept free from battle.

Third, and ultimately most important: Muslim holy days, including the fasting month of Ramadan, were scheduled entirely on a lunar basis, which means they varied, and continued to vary, in their seasonal occurrence from one year to the next – each festival almost two weeks earlier (in solar terms) than its counterpart from the previous year.

For non-religious purposes, Islamic regions continued to add a solar adjustment, so that the agricultural cycles, for example, would mirror the seasons. Most political scheduling also occurred according to this lunar–solar combination. Several regional systems were used within the Islamic world, ultimately of course including the very precise Persian solar year.

The Islamic religious calendar encouraged a great deal of astronomical research and observation in order to determine the precise day at which, in the lunar cycle, a religious observance would begin, usually at the first sighting of the crescent moon in the appropriate month (the ninth month of the year, for Ramadan). Some disparities in specifics resulted and continue to result from, among other things, the precise location of the calculation: a crescent moon's arrival can vary by a day or so, depending on place of observation. Some efforts to schedule major rituals by the sighting in Mecca have usually been downplayed in favor of local sightings.

The Net Results

By the 16th century, with the Gregorian calendar gaining ground in Europe, most of the basic calendar work was done, in most of the major civilization areas – though as we have seen the Chinese continued to tinker. Further astronomical observations mainly added small details, not basic reassessments.

This said, the most obvious feature of the human commitment to calendars by this point was the sheer variety involved, even though each system was based on careful scientific research and a common need to organize time, in a given year and in years past and future. Extremes ranged, clearly, from the pure lunar to the pure solar, with a lunar–solar combination of some sort the most common result. The variety meant that months were measured in different lengths depending on the system, while new years started at different seasons – the Roman and then Christian decision for January 1 was actually unusual. In the world as of the 15th century, the problem posed by having so many different regional calendars was not yet acutely felt.

An ongoing legacy of this long period of calendar development, aside from some continued variety, involves the impressive number of religious or quasi-religious ceremonies that are still scheduled according to a partially lunar (or, in the case of Islam, entirely lunar) basis. Christmas, to be sure, has a fixed date, after several centuries of debate among believers as to when Christ had been born. The ultimate decision, emerging in the 4th century in Western Europe, to fix on December 25 presumably resulted from the date's correspondence to the older Roman winter solstice celebration on the same date – though a charming Christian story held that Christ needed to be born on the shortest day of the year (which the Roman calendar at that point listed as December 25) so that steadily increasing light would accompany his growth.

But most of the rest of the classic holidays (in contrast to modern political holidays, which all have a fixed calendar date) rest on a lunar base modified by a solar correction (outside of Islam) to keep them in the proper season. Thus Easter is set for the first full moon after the March solstice (the 21st). The Chinese New Year, or spring festival, begins on the new moon that appears between January 21 and February 20 – like Easter, the specifics varying each year through the lunar cycle. (New years in other parts of East Asia are calculated similarly, though sometimes with slight variation.) Jewish ritual days are also based on the modified lunar approach: the Hebrew calendar, with lunar months, is brought into line by adding a short month seven times over a 19-year span. A ceremony like Passover (scheduled for the 15th of the month of Nisan, the seventh lunar month of the Jewish year) will thus always occur in spring, but at varying solar dates. The Hindu harvest festival, Diwali, coincides with the darkest night of the lunisolar month Kartika – usually between mid-October and mid-November, but again with variable annual specifics. And the list could be expanded. Many contemporary cultures, in other words, though increasingly wedded to the need for greater uniformities and far from a traditional fascination with the moon, cherish the complexities of an older approach to ordering time when it comes to some of the most meaningful celebrations of the year.

Measuring the Time of Day

Interest in improving the calculation of time during the day was at least as great as the ongoing concern with calendars. Here too there was ample inheritance from the past, but also many opportunities for greater precision. In some cases, at least, the steady interest in more precise measurement may have had more impact on ordinary life – aside from the scheduling of major festivals – than the calendrical ruminations of the astronomers did.

Several lines of developments were important. The clearest trajectory involved continued work on timepieces themselves, particularly the water clock but also other instruments to measure duration. While there were no revolutionary breakthroughs as yet, there were some fascinating investments and, at the end of the period in Europe, even the crafting of the first mechanical clock.

A second set of innovations occurred at the scientific level, where the traditional, somewhat casual approach to daily units was modified in significant ways – though some calendars, like the Hebrew, continued to be fairly vague in defining the day beyond sunset (when a new day began) and sunrise. While great variety persisted, the notion that the day had 24 hours, and even that each hour had the same length, gained ground. Ultimately, definitions of hours meshed with some of the new work on instrumentation, though various kinds of imprecision persisted.

Finally, new ideas about daily time (units of measurement and timepieces alike) began to develop new intersections with some aspects of ordinary life, particularly of course in the cities. In this period the most important changes in the way daily time was handled occurred in the religious domain, where for many people some fundamental shifts occurred. But this could blend with other requirements, particularly in the activities of merchants, which would set the stage for other innovations later on. As with calendars, however, key changes and even felt needs varied regionally, setting up some important issues for later analysis as well.

Recording Daily Time

In 807 CE the great Islamic caliph, Harun al- Rashid, sent the European emperor Charlemagne a gift of a magnificent water clock as a gesture of good will. (He also sent an Asian elephant.) On the clock's dial were 12 small doors. A new door opened at every new hour (activated by the water flow) and, from it, one or more little balls fell on a drum – their number corresponding to the hour. And then, at the end of the day, on the 12th hour, 12 miniature horsemen appeared and shut all the doors – preparing for another 12-hour cycle. The members of Charlemagne's court were properly awed by this elaborate device, which at that point they lacked the skill to emulate. The whole exchange, apart from revealing a technological disparity, showed the fascination of the ruling classes, in virtually every civilization, with challenge of keeping time.

Innovations and elaborations occurred throughout Asia and Europe, from the early classical period onward (see Figure 3.1). As the caliph's gift suggests, Arab engineers devoted much attention to improving clock technology. One engineer in the late 8th century, Al-Jarzari, working for a regional king, made various clocks in different shapes and sizes; a book devoted to his work illustrated 50 mechanical devices, including many water clocks.

Elaborate structures were created in a number of major civilizations. In Athens, as early as the 1st century BCE, an astronomer supervised the construction of what was called the Tower of the Winds. This was an octagonal structure combining an elaborate set of eight sundials, a water clock, and a weather vane (for determining wind direction) – making this an early example of a clock tower. The building, now restored, can still be visited. Roman engineers also worked on clocks, including efforts to improve the mechanisms and accuracy of water clocks.

Water clocks spread widely in Christian Europe in later centuries. Many major churches had one of their own. In the late 12th century, there is a report of a fire in one English abbey church, where the monks "ran to the clock" to get the water to put out the blaze. By this point also in English the word "clock" was coming into wide usage, replacing the earlier term horloge, which had come from the Greeks; clocks was a Celtic word, initially meaning bell.

The Byzantine Empire also harbored elaborate water clocks. There were at least two big public clocks in Constantinople, one linked to the great St. Sophia cathedral. Another clock in Palestine (before Arab conquest) featured two dials, each denoting six hours; as an hour struck, figures such as Hercules and Medusa paraded out on a cogged wheel.

Chinese societies probably put the most consistent effort into clock development before the 13th century. Emperors sponsored a great deal of research, while also

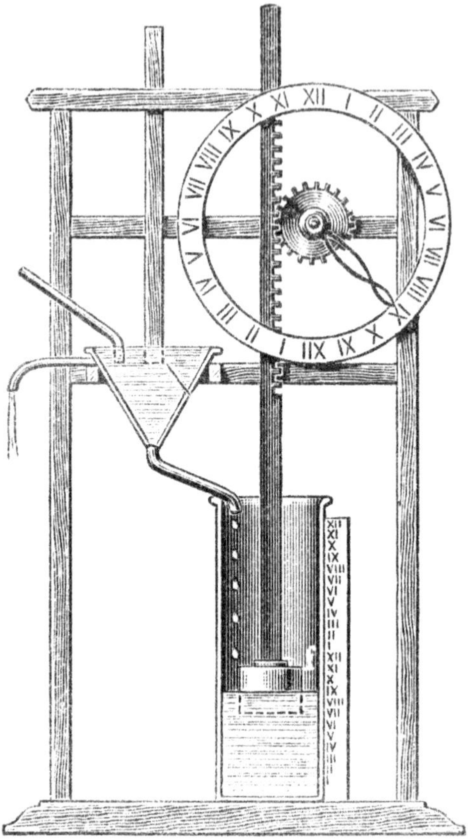

Figure 3.1 Vintage 19th century old drawing of a Egyptian water clock. While the date of this clock is uncertain, it is recent enough to have used Roman numerals, though the clock face represents an older notion of dividing the hours of the day. Picture credit: Oprea Nicolae / Alamy Stock Photo

trying to keep careful control over the officials charged with determining the time, among other things limiting their communication with other scientists.

Particular attention was given to improving the water clock. In 725 CE, Yi Xing and Liang Lingzen created a water-powered cogwheel clock. By this point, the most elaborate water clocks went well beyond simply pouring out measured amounts of water, allowing time to be determined by recording the rising levels in a receptacle (which in turn was initially calibrated by using measurements on a sundial). The goal now was to create mechanisms that would use water power – still in measured amounts – to move a dial, that in turn would indicate the time. The new Chinese device had this feature, through use of the cogwheel. In the 11th century another scientific genius, Su Song, used this kind of clock to construct a huge astronomical clock tower, in 1088. This featured a rotating armature that would move the dial, powered by the measured flow of water in three of the year's seasons and by liquid mercury in winter (when mere water might freeze). There is evidence that Arabs also used mercury, and transmitted the idea to Europe. Trade exchanges continued to provide opportunities for the major Asian and European societies to learn from each other.

A crucial advance, probably widely introduced in Europe only in the 13th century, applied what is called an escapement mechanism to the water clock. This device controls the movement of the gear that in turn moves the clock's hands. Each impulse, for example from a water flow, releases a gear tooth that advances the hands of the dial at a steady rate. Escapement mechanisms would be fundamental in allowing the development of mechanical clocks, but their initial invention undoubtedly constituted just a further step in water clock improvement.

Initial work on escapements dated back to classical Greece. An engineer named Philo of Byzantium wrote of a counterweighted spoon, activated by a steady water flow, that emptied regularly but was then pulled up by the counterweight. In China a Buddhist monk, working with a government official, introduced an escapement into a water-powered device. Escapements were then used in the elaborate water towers built during the Song dynasty, in the 11th century. In Su Song's tower, escapements tipped the water container over each time it filled up, advancing the clock's wheels according to a regular rate of flow. Knowledge of this technology may have spread from China first to the Middle East, through the medium of merchants, and then to Europe. A Spanish book on clocks sponsored by a 13th-century king referred to escapements activated by liquid mercury, copying technology available in the Arab Middle East.

Water clocks offered a mixture of advantages and disadvantages. Most obviously, they allowed calculation of time during the night or on cloudy days. One of the reasons for their popularity in Christian Europe undoubtedly depended on the need for timing despite frequent rainy weather. In many societies also – China most obviously, but also the Middle East and Europe – the big water clock towers allowed a mixture of daily timing and astrological calculations. Moving dials recorded time, to be sure, but they also represented figures of the moon and key stars, their motion connected to astrological forecasts.

But there were drawbacks. Pure water clocks, obviously, had trouble during winter in the more northern latitudes because of freezing, though the introduction of mercury use could help here. In the best of circumstances, they were not terribly accurate. It was hard to regulate flows precisely, and changes in temperature (even aside from freezing) could alter rates. Almost certainly, most water clocks had to be readjusted every few days, through the juxtaposition of sundials, to keep them roughly on time.

The clocks were, also, labor intensive. A human attendant had to oversee them constantly, because so many things could go wrong with the water flow. This meant, in turn, that the clocks were expensive not just to build but to operate, which in turn confined their use to the wealthier religious establishments or royal courts. These were not, for the most part, devices available to a wide population, even in the cities.

Finally the most elaborate clocks – the most accurate, though also sporting the fanciest mechanisms – were extremely heavy and large. Su Sung's clock tower in

China, for example, was 40 feet high. The clocks were irrelevant, as a result, for people on the move. They could not be used on ships, where the tossing of waves made any sort of regulated flow impossible. The advance of water clocks, in sum, showed the deep interest in feeding the human concern for time. For many centuries, they offered the best available, day-round option. And for many societies, as their mechanisms improved, they might seem quite adequate, given the fairly loose needs for daily timing that long persisted. But their limitations might spark a desire for alternatives.

Technological advances also applied to other timing devices, besides the water wheel. In China and elsewhere, the use of calibrated candles to tell time became more accurate, from the 6th century CE onward. Candles were carefully engraved with evenly spaced markings. This was particularly useful for people who wanted to know how much time had passed since they went to bed. Whether separately invented or learned from Asia, candle clocks were popular with English kings by the 10th century.

The spread of Buddhism brought wide use of incense candles, which again could be calibrated according to precise units of time. These were widely used to regulate some devotional practices in Buddhist monasteries (particularly after sunset), but they could affect other activities as well. A Chinese poet celebrated the utility of incense candles in the 5th century: "By burning incense we know the o'clock of the night. With graduated candles we confirm the tally of the [night] watches."

Oil or incense lamps offered a similar option. Again in China, improvements were introduced in the shaping of the ceramic lamp, to make it easier to determine the rate of oil used. The lamps were less likely than candles to blow out, but, obviously, like candles they measured the passage of time from one point to another; they did not indicate the time of day. The same applied to the continued use of sand glasses.

And this leads to a final puzzle, in the history of daily timing until the 14th century: it was not always clear, even for the people most interested in, and closest to, the available timing devices, how much concern existed for knowing what time it was. Some of the water clocks, for example, had no faces. They were used to activate a bell or some other device at particular intervals, most obviously to indicate when it was time to start a particular religious ceremony; but they did not make it easy to determine time otherwise. When clock faces did exist, they often indicated only hours, or possibly quarter-hours – they were not further divided. Quite apart from limitations of technology, the nature of needs for time continued to vary and were often surprisingly imprecise.

Determining the Hour

Over the centuries, from the classical civilizations onward, there was increasing motivation to agree on divisions of daily time, a subject that, as we have seen, was already frequently discussed in the early civilizations. The first motivation was scientific or mathematical: a more regular calculation of time units during the day would promote a sense of order and would facilitate other calculations. The second motivation was practical: as timepieces improved – and, particularly, as it became possible to move a device over a dial on the water clock – it would certainly make sense to agree on some regular measurements on the dial itself – even if there was not, yet, a widely felt need to be able to tell the time apart from specific ceremonies. Even sundials could prompt an interest in regular markings.

The process of regularization was, nevertheless, slow and uneven. Roman society remained largely interested in measuring time during the daylight hours; night could be divided up more crudely. Many localities, into the 14th century and beyond, continued the old practice of timing hours variously depending on the season – longer hours in summer, shorter in winter. This pattern appealed to the old notion of dividing the day into two 12-hour spans, regardless of season, and it obviously could be made to work if one knew the local tradition. For many people, it was far more important to use the natural division of night and day than to develop a more uniform basis for calculation regardless of season.

At the level of theory, by the 5th century BCE a Greek scientist advanced the notion of dividing the whole day into 24 units, again for purposes of mathematical calculation more than as a practical gesture in daily timekeeping. Hipparchus is credited with the idea, in the 2nd century BCE, which applied the old Mesopotamian interest in using units of 12 – previously used in the measurement of circles – to this timing issue. His definition referred to the even division between night and day on the spring and autumnal equinox – when 12 hours of day and 12 hours of night might literally apply, but now extended the units to measurements in other seasons.

The idea of a 24-hour day also emerged in China, though we have seen that early Chinese scientists also used divisions by 100. Scientists in the early empires often split the day into units of 12 – i.e., two-hour units by our standards. In this calculation, the day began at 11 PM, with the day's first unit running from then until 1 AM. But under the Tang dynasty this was refined into 24 hours.

On a more practical level, hours began to be used for marking timepieces possibly as early as ancient Egypt. The obvious need was for some accepted system for sundials, beyond simply calculating the angle of a shadow. Egyptians saw the day in terms of 12 units, possibly accepting this kind of calculation from their Mesopotamian neighbor, possibly taking 12 over from the annual number of lunar cycles. This did not initially mean a commitment to seeing the whole day in terms of 24 units – after all, sundials did not work at night; and we have seen that even Greeks and Romans, heavy sundial users, did not bother with dividing up the night very precisely.

When water clocks had dials, they were not usually divided up into 24 units – though the notion was available. In Europe, many dials were set according to an older Roman pattern of dividing the day into three-hour units – used to mark time in the Roman marketplaces – which was adapted into Christian ritual. Bells rang for the first worship period, prime, around 6 AM, terce around 9, sext at noon, nones at three, and finally vespers at 6 (or, more roughly, at sunset).

And even when mechanical clocks were first introduced in Europe their dials were initially adjusted to show units of the day and those of the night with seasonably varying lengths – which meant that the length of their pendulums had to be altered literally every day. Fairly soon, however, this nuisance or nicety was abandoned in favor of a straight 24-hour cycle, running from one to 12 twice each day but with hours of even length throughout.

Finally, one other measurement change came into view at the about the same point that mechanical clocks began to emerge – though there was probably no initial connection. The first known reference to the "minute" as a unit within an hour comes from a mathematical treatise by John of Sacrobosco, around 1235 CE. The treatise obviously took advantage of the old Mesopotamian interest in divisions of 60, again applied to measuring circles – it had no correspondence to any natural phenomenon. In 1267, the English scientist Roger Bacon talked of the division of time between full moons in terms of the number of hours, minutes, seconds (and he also cited thirds and fourths). The notion of actually using seconds would not be picked up for many centuries, because there was no need for them and no way of measuring them. But the notion of hourly divisions by minutes might begin to make some sense as the dials of water clocks and then mechanical clocks became more precise.

This evolution toward thinking in terms of hours and possibly minutes was obviously a complicated one, ultimately linking rather abstract mathematical interests to practical capacities to calculate more precise intervals and note them on timepieces. Popular traditions, content with looser or more variable systems, long persisted. Even when it became technologically possible, thanks to improved escapements, to record smaller units on clock dials, this was not widely or quickly put into play. Clock dials long continued simply to register hours, or at most to divide the hours themselves into halves and quarters. There was not much need for more – though by the 14th century the capacity to offer more was beginning to emerge. Unevenly, acceptance of the standardization of the hour did advance, in many regions, which was an essential preliminary to any interest in units within the hour.

Daily Uses of Time

Not surprisingly, actual uses of time are harder to track historically than the development of timing ideas and mechanisms. Even employment of calendars varied, apart from increasing agreement on some religious markers like Ramadan or Easter. Daily time is even more elusive.

Most people, even in the cities, did not have easy visual access to a timing device, though, of course, they could and did use their eyes to note the position of the sun. A few public water clocks were established, as in Athens or imperial China, and sundials were available in many urban centers. Major religious structures often had water clocks. But for the bulk of the population, a casual opportunity to check the time was often not readily available.

There was one important exception. In some regions dependent on irrigation, as in Mesopotamia, timing devices became crucial in indicating start and stop times for each individual taking water from the system. Hour glasses were widely deployed for this purpose. But while precision was essential, it did not depend on scheduling in terms of time of day. Here, as needs developed, they applied primarily in the cities, where getting people together at appointed hours might become more important.

In this context, the opportunity to convey time by sound rather than sight became vitally important, particularly when one of the key uses of time centered on gathering people from a considerable distance within a city (even granting the smaller, more compact size of most cities before modern times). Several devices were available, as we will see. The human voice was one, vital in conveying time in Islam. Uses of horns – like the ram's horn traditionally used to call to Jewish services – or drums were other options, developed early in the human experience.

But ultimately the most important invention was the bell. Bells were probably invented in China, by hunting and gathering groups. Made of ceramics, they were devoted to music. Metal bells have been found as far back as 2000 BCE, in Asia and elsewhere. They were probably used not only for music, but also to track movement – attached to dog collars and ox yokes for example, or chariots. Small bells also served as ornamentation, on priestly robes. Obviously, however, as metal technology improved and somewhat larger bells could be constructed, they were also ideal for indicating time.

Thus the Romans sounded a bell to indicate the opening of the public market, and again to warn of its close. Bells were also used to announce the opening of the baths, that vital Roman leisure center. We do not know whether strict timing was involved here, as opposed to daily improvisation. But it is important to remember that Roman markets and baths spread widely around the Mediterranean and into northern Europe, exposing a variety of people to the idea of marking urban time for some regular services, and using bells in the process.

Many urban centers also used bells to alert to the closing of the city gates or to indicate the beginning of nighttime curfews. Chinese cities under the Tang dynasty used trumpets and drums to announce the closing of city gates. None of this was necessarily time-sensitive –gate bells or trumpets could be sounded according to the position of the sun. But the capacity for a more explicit connection to timing clearly existed.

Urban developments encouraged some other uses of time, though not necessarily with bells. Legislative assemblies met regularly in Athens for example, in principle inviting the participation of all citizens. They were normally scheduled on four set times in the annual calendar – thus assuming one use of time; there is some debate as to whether other meetings might be called, and we do not know how the daily opening time of a session was determined or marked.

Time use has a clearer evolution in the history of the Roman courts of law, another vital political institution in this particular society. Legal procedures, and even the use of lawyers, evolved slowly in republican Rome. But by the later Republic there are accounts of lawyers busily scurrying to be on time for the opening of a court session; lawyering was extremely competitive, so timing may have been crucial – probably determined by consulting a sundial in a public square, near where trials themselves were conducted. Anxiety about scheduling might have prompted the kind of lament about time noted at the beginning of this chapter.

Roman courts certainly prompted another time concern. For a long time, lawyers' speeches were not regulated. In the trial of one Bassus, in the 1st century, the problem became obvious. Bassus was an important figure, and lawyers on both sides were eager to participate. Two prosecutors spoke for several hours. Then Pliny, for the defense, countered with a five-hour diatribe. Then two more orations took what were simply the preliminary procedures into a second day – before any evidence was presented. Small wonder that Roman judges increasingly used a water glass to regulate the length of speeches, deciding the size of the glass – and therefore the time available – by the nature of the case.

Concern about time's duration – though not yet clock time – applied to other settings also. As several major religious developed more elaborate theology, and less strict dependence on ritual, religious services in some cases included growing space for sermons. These, in turn, could drone on. In response, at least in some Christian churches, hour glasses were used to suggest limits; the only problem was, when an ardent pastor simply flipped the glass to extend the duration, as often happened – who was in a position to object?

All this, admittedly, is varied evidence of the actual uses of time, scattered in chronology and place alike. We are far away from systematic generalizations about the ways time was allocated and perceived in the cities, though it is clear that urbanites had a considerably different time experience from their rural counterparts. Urban needs and devices probably did not yet impose a deep consciousness of daily time save for the members of a few professions – but they did reflect changing needs and the growing availability of timing devices as well.

Into this general pattern, the requirements of at least two dynamic religions contributed significant, and measurable, innovation from their early days. Both Islam and Christianity helped shape the nature and uses of the calendar, as we have seen. They may have had even more important results in their demands on daily time.

The link between religion and daily time was not new, of course. Egyptian priests and others had developed a careful timing of rituals, as part of the regular honor due to the gods. But they made no attempt to involve a large number of people in this time sense, and their schedule depended largely on movements of the sun – most particularly sunrise. Jewish services carried the use of timing further, in trying to set up several points in each day for prayer. As we will see, both Islam and Christianity built on these precedents, but went further.

Buddhism, the other great missionary religion, developed a different take. The religion was not heavily reliant on ritual – this was one of its objections to Hinduism. The strong emphasis on meditation, for ordinary Buddhists and monks alike, also moved away from a rigorous time sense. Some Buddhist thinkers, particularly in the Theravada tradition, argued more fundamentally that time was not an objective reality but a projection of human consciousness: "Time…does not exist by nature, it is merely a concept."

This does not mean that Buddhists systematically rejected or ignored time. Buddhist monasteries came to develop something of a daily schedule, including periods for meditation, though it was not terribly detailed. Individual Buddhist scholars, as we have seen in China, often contributed directly to the generation of improvements in the technology of water clocks. But Buddhism – regionally varied in any event – did not mark a major stage in the elaboration of either daily or calendar time, with the partial exception of its role in disseminating the use of time-measured incense candles.

Islam

Islam quickly developed a crucial innovation in daily timing, even as Muhammad explicitly moved to link the lunar calendar with religious prescription. From the start, the Prophet urged a series of prayer times during the day, to which all the faithful were supposed to respond – the obligation was one of the five core requirements of the new religion. Initially Muhammad may have had three times in mind, but this ultimately turned into five. The faithful Muslim was to pray at set points in the day: at dawn before sunrise; at midday, after the sun passes its highest point; in the late part of the afternoon; just after sunset; and between sunset and midnight.

Some of these times could of course be determined by sight, but others required some kind of timing device, at least in principle. The result, in the early days of the religion, was an obvious challenge. As the number of faithful grew, and in a population that lacked wide access to clocks, a decision had to be made about how to alert the group at the appropriate times. Muhammad presumably considered the methods currently in use by Jews and Christians (ram's horns for the former, wooden claques at that point for the latter), but decided on another method: designating an individual with a pleasing and powerful voice to issue the call (in fact, usually two calls: first to remind that the time had arrived, second to note the beginning of a service). This would provide coordination and also allow the caller – known as the *muezzin* – regularly to remind the faithful of the basic tenets of the new religion, beginning with a fourfold assertion of the greatness of Allah.

The Prophet's first choice for the *muezzin* was Bilal ibn Rabah, a former slave, son of an Arab father and Ethiopian mother, who was an early convert and had a powerful voice. Bilal and his initial successors simply issued his call from a rooftop. But by the later 7th century, Muslims began to add a slender tower to their mosques, the minaret, from which the call could be issued. Since in principle enough mosques should be built so every neighborhood was with range of the call, the system provided ample coordination. Many mosques presumably had access to the standard combination of sun dial and water clock. But it is important not to exaggerate. Descriptions of the daily call frequently referred more to sighting of the sun – "when the sun's shadow begins to wander toward the east" – rather than to any artificial device. Doing the right number of prayers, in the right daily sequence, was more important (and feasible) than insisting on timing precision.

Furthermore, popular responses surely varied. It was perfectly acceptable to pray individually, facing in the direction of Mecca, or to join a service in the mosque. But whatever the individual reaction, the daily calls to prayer set the rhythm of the day as a whole, at least in the cities (Islam spread more slowly to the countryside). And this remains the case today, though now the calls are issued over loudspeakers. Here, clearly, was a crucial innovation in the routine exposure to daily time.

Christianity

For most people, the impact of Christianity on a daily sense of time was not too different from that of Islam. Like Islam, Christianity tried to promote a more active daily commitment to the worship of God, with a schedule attached. The specific timing varied, and the alerting devices differed – bells instead of tower-top calls. But the ideas were similar, and the impact on ordinary life might have been similar as well – though the Islamic appeals for participation were more urgent. In both cases, rural patterns were far less affected than urban.

However, Christianity, as it matured particularly in Western Europe, did develop some religious schedules that were even more precisely regulated by the clock. Only a small minority of people were directly involved, but it is possible that, ultimately, the sense of time had wider results, particularly in associating diligent work with some sense of the clock.

Christians did not go as far as Islam in dividing the day into points of religious obligation for the faithful in general. But from the early days of the faith, religious officials (and those aspiring to the priesthood) were called upon to pray at set times during the day. This built on an older Jewish practice of designating four specific hours for prayer. Called the Liturgy of the Hours, it involved an official set of prayers marking the hours of each day and sanctifying the day with prayer. While ordinary faithful did not need to follow this rigorous schedule, they were encouraged to be aware of it and participate at least in part.

Schedules evolved during the early centuries of Christianity. Jewish traditions of prayer times were combined with Roman practices for organizing the business day – often including the use of bells, at least within a church, to designate timing. (The addition of exterior bells, in church steeples, only began to take shape from the 5th century onward, and of course they were expensive.) Beginning with prime, at six in the morning, prayer services and recitations were conducted at four other points (9 AM, noon, 3 PM and 6). In the early church, various miracles and conversions were associated with several of these times.

Again, this detailed timing did not necessarily apply to ordinary Christians. But they were encouraged to join in at least some of the scheduled services. And as the physical structures of churches became more elaborate, particularly in the cities, church bells sounded out the canonical hours, impressing a sense of daily time on the wider population whatever their religious commitment. As with Islam, though with different specifics, religion was fostering a new sense of the clock. But, as with Islam as well, the recommended timing for each service should not be taken too literally. It was the order of the services, not the hour, that really mattered, and over the centuries the specifics changed as well, in some cases creating major confusion during the transitions.

Yet in fact, as early as 606 CE, the Catholic pope urged Christian churches everywhere to organize water clocks and sundials, "to distinguish the hours of the day". Many of the devices were crude, but the linkage between clock time and religious time was clearly advancing.

Further, one vital group in Western Christianity carried the notion of religious scheduling to greater extremes. Christianity, early on, had motivated a number of individuals to devote themselves to a life of prayer and sacrifice, sometimes in isolation. And while church authorities welcomed their fervor, and their pious example, the proliferation of religious hermits did pose a problem, for some could develop unorthodox ideas and practices that might unsettle established doctrine. In response, the church ultimately encouraged a more organized framework for this kind of religious commitment – and a careful sense of time was a crucial component of this sense of structure.

The great innovation stemmed from Benedict of Nursia, who died in 547 CE. Benedict founded a number of communities for Catholic monks – ultimately leading to a whole monastic order – and, even more important, developed regulations for their conduct. In the monastic world of the Benedictines, not only prayer but also work was to be rigorously scheduled on a daily basis.

Thus the day was divided into carefully timed segments, including, of course, the canonical hours. It began at midnight, with prayers; followed by another service at 3 AM. These two services were usually held in the dark, until better candles became available in the 14th century. Monks were expected to sleep in the intervals between the services, before rising again at 6 in the morning for another reading, followed by work or prayer until 9, when high mass occurred. Another service occurred at noon, then a period for rest until 3. Farming and housekeeping work then occurred until sunset, when there was another set of prayers; the final night prayer, or Compline, took place at 9.

All of this, with the obvious exception of seasonal allowance for sunset, was quite carefully timed in principle. Some scholars, headed by the great German sociologist Werner Sombart, have seen in this monastic commitment a major step forward to the kind of rational organization of time that, under a very different framework, would ultimately generate the capitalist approach to time management. Another historian, Evitiar Zerubavel, makes the bold claim that "the Benedictines were clearly the first ones to have established a rigorous temporal regularity that was not directly geared in any way to nature". Yet here too, what was on paper, and what actually happened, might easily vary in the absence of precise timing devices. When monks were late to a morning service, for example, the service was often simply shortened; time-based punctuality was not pressed too far. The main point was to get all the services in during the day, not to insist on their timing or duration.

Benedict's rule was unquestionably widely influential, shaping virtually all the major monastic and (women's) conventual orders in Western Christianity from that point onward – orders that, in turn, had a substantial impact on the religion as a whole, sense of time potentially included. It is really impossible to know whether some Europeans began at this point to develop a novel idea that units of time had to be accounted for usefully, and not simply applied to religious services. But it was at least possible, and we know that a variety of urban activities began to be scheduled around the sounding of the church bells.

And one result was very clear. Some Christian monasteries became very interested in improving their instruments for determining daily time, not content with primitive water clocks and sundials. Thus, an Italian archbishop in the 9th century developed a new observation tube to check the night sky, claiming that it allowed time to be determined after dark "even without the crow of the cock". Thus, a monastery in Catalonia (northwestern Spain) in the 10th century deliberately reached out to Arabs in the Muslim-controlled city of Toledo, seeking instructions on building a more advanced water clock. Monasticism could thus supplement the more general Christian interest in timepieces and their role in proper religious life.

Mechanical Clocks

One other striking, though initially very primitive, innovation did emerge from West European cities by the 13th century: the first purely mechanical clocks, no longer dependent on water flows with their vagaries and labor requirements.

Here was the setting. By the 11th century, European cities, though puny by Asian standards, were growing. Manufacturing and trade were expanding. Churches dominated the urban landscape, their bells clearly marking key times during the day. Their bells might, of course, regulate more than religious services; for example a tradition arose that set the scheduling of duels for when the church bells signaled the afternoon service. Bells for the evening service might be used to mark the end of the workday on the eve of a religious holiday. But cities also maintained bells, to help merchants and (in the ports) navigators determine the local schedule; these bells normally sounded at noon. Each city had its own particular schedule of hours, apart from the Christian framework: the problem of local variability may have been particularly great in decentralized Europe, most obviously compared to China. Each locality preferred to define hours explicitly to reflect its own particular longitude and therefore its distinctive relationship to the positions of the sun. (Decisions about how to set the daily clock would also reflect the season in which choices were made.) Noon in Basel, Switzerland, thus occurred at 1 o'clock, according to the patterns used by some of the other Swiss cities. The result was a clear need to clarify local time for business and navigation purposes, whenever one went from one city to another. Town bells could help a traveler figure out the local system. But the larger point, that many cities were developing a greater interest in defining daily time, for government and commercial purposes, remains valid. Along with the religious interest in timing precision, and the drawbacks of having to rely too much on water clocks, there was obvious potential interest in further innovation. Finally, advances in clockmaking in other societies might serve as stimulus in Europe, as transregional contacts expanded. Thus a 13th-century Spanish translation of an Arabic work suggested possible invention of a mechanical clock (this is not certain), while Ibn Khalaf al-Muradi definitely improved on a mercury-driven cogwheel clock.

Exactly when the first mechanical clock emerged in Europe is not certain – some claims go back to the 11th century. References to clocks and horloges in church writings accelerate from the late 13th century onward, more clearly suggesting both growing interest and probable innovation. The key development was the use of gravity-powered pendulums to replace flowing water or mercury in activating the gears that in turn drove a dial mechanism, the rate of pendulum descent controlled by the escapement device.

The early mechanical clocks could serve two purposes: they could model the movements in the solar system while also providing time signals for the schedule of events or other activities. Early clocks were often installed in towers. Some had no clock face, but simply activated bells that would indicate the timing of religious services. When dials were available, they often had only one hand – because knowing the hour was precise enough. But, obviously, further elaborations could be added over time.

In 1283 a large clock was installed in Dunstable Priory, in England, and soon after a "great horloge" was set up in Canterbury Cathedral. Over the next 30 years many churches in monasteries in England, France, and Italy began to mention clocks, often noting that the new devices replaced water clocks. In 1322, in Norwich, England, a large new clock was installed, at great expense, complete with bells to toll the hours; two skilled clockkeepers had to be employed for two years to oversee the operation.

The new clocks were not usually very precise. Though they required less labor than a good water clock, they had to be wound every four hours. But, without question, the new clocks set the stage for a major surge in the uses of time, first in Europe and then more widely – marking the end of a long and diverse period of development and inaugurating a new era.

Conclusion: Some Conundrums

The Comparative Issue

European gains in timekeeping by the 14th century were in some ways surprising, and they raise some challenges to interpretation that will also spill over into the next period, covered in the following chapter. The European economy was gaining ground by this point, after the long doldrums following the fall of the Roman Empire. But Europe remained decidedly backward by the standards of the great Asian societies, economically, politically, and technologically. To be sure, by the 14th-century Europeans were beginning to adapt inventions they had learned about from China and the Middle East – the compass, gunpowder, paper manufacturing, and, soon, printing. But the development of the mechanical clock may have owed less to copying from the techniques of more advanced societies – though models from the Islamic Middle East played a role. Why was Europe poised to leapfrog over other societies when it came to keeping daily time, and also to define the components of the hour more clearly in principle?

It is important not to overdo. Mechanical clocks borrowed many features from the advanced water clocks that several societies had developed – particularly the gear arrangements that allowed a pointer to pass over the dial. And by the 14th century Europe's rudimentary mechanical clocks did not yet keep time as well as the great water clocks of China and the Middle East. Still, we can see in retrospect the makings of a major step forward, beyond the achievements of the great Asian civilizations.

Several factors may have been involved. Some historians have emphasized the freewheeling atmosphere of European cities, where merchants enjoyed great power and artisans were able to tinker with mechanical improvements. They contrast this with the situation in China, where timekeeping was more rigorously controlled by a bureaucratic state that did not encourage as much communication or innovation from below. This is certainly worth considering, but it is vital to remember that innovative Chinese technology had long led the world in other areas, and that their water clocks were quite advanced.

Another approach, admittedly less common among the admirers of the Western technological spirit, would go back to the role of Western Christianity in emphasizing the importance of keeping reasonably accurate daily time, particularly of course in the monastic movement. The ubiquity of great church bells in the cities, marking time among other functions, added to this religious climate. Chinese Buddhists were interested in time as well, and contributed some important ideas, but the importance of scheduling was less extreme.

A third approach would emphasize the new needs that were developing as European commerce expanded. We have seen that many urban merchants tried to keep track of time. Add to this, in a society where seagoing commerce was gaining in importance, the relevance of timekeeping to navigational devices.

Most important is the likelihood that these factors combined in some fashion. European merchants were not uniquely skilled but they may have gained disproportionate time-consciousness from Catholic religious precedent. Religious interests could encourage artisan inventors, who might see their work as contributing to pious devotion as well as urban life more generally. It is of course impossible to pinpoint the factors involved, or to explain the comparative dilemma definitively.

It is also vital to remember that European water clocks, though highly developed, were inferior to their Asian counterparts, which may have made the search for alternative devices more important. In the Middle East, further, water clocks and sundials did not have as many drawbacks as they did in Europe: seasonal changes did not have much impact on the visibility of the sun (and there were fewer cloudy days overall), and there was little or no freezing to disrupt the operation of water flow. (This may also help explain why northern Europeans generated more innovation in this area than southerners did.) Again, Europeans, at least in places like rainy England and the Low Countries, may have had some distinctive problems and needs.

Religious and Merchant Time

A second challenge to interpretation, within the European context itself, can be stated more quickly, though it may relate to the comparative dilemma. Several historians, particularly in the great French tradition of social history, have argued that, by the end of the period considered in this chapter, Europeans – urban Europeans at least – were beginning to shift from religious interest in time to the rather different concerns about time manifested by merchants (and navigators).

We have seen that, through most of the long period covered in this chapter, religious considerations, sometimes combined with the concerns of the more bureaucratic states, shaped most of the interest in daily time. There is no question that the role of religion and ritual in the history of time deserves major emphasis.

Ultimately, however, the religious interest in time would be overshadowed, though not eliminated, by more commercial considerations. Is it premature to see this transition taking shape in 13th and 14th-century Europe, rather than later on? Certainly for most people in the cities, awareness of time was still mainly conveyed by church bells; mechanical clocks were primitive and of limited availability. Further, the first public installation of mechanical clocks usually occurred in cathedral towers, as in the several English examples. And in the countryside, where the vast majority of Europeans still lived, time still usually meant calculations based primarily on nature supplemented a bit by local church rituals.

Both of these issues – the comparative position of Europe and the shift toward more commercial calculations – would become clearer in the next period in the history of time, during and after the 1400s. Here is when European innovation became far better defined, and the implications of time for daily life began to extend. Here is when, further, questions begin to arise about how a European lead in clocks and timekeeping may have spilled over into other innovations, in what turned out to be a ramp-up toward the industrial revolution.

Before racing ahead, and before risking overemphasis on a European distinction, a bit of stock taking is in order. From the emergence of the more elaborate civilizations of the classical period to the flowering of missionary religions, timekeeping evolved significantly – again, without truly revolutionary breakthroughs. Both calendars and daily timing saw major changes, in concept and technology alike. Fascination with time tended to shift increasingly from daily contemplation of nature – though this of course continued – to more elaborate calculations.

Several strands combined. A desire to impose rational structure on time, shared by many scientists and some political leaders, contributed to a number of important changes. The development of more elaborate ritual needs, and their impact on the way years were calculated and days conceived, were possibly even more important. And practical considerations, themselves evolving with the growth of cities, added a third component. Practical and religious needs, combined, help explain why time was increasingly connected to sound – the bells, the calls to prayer – a linkage that continues in many ways today.

And all of this, in regional societies at best loosely connected, operated in an atmosphere of great variety. Lunars and solar combinations continued to develop in different ways. Definitions of hours, though beginning to regularize a bit, continued to fluctuate, as did definitions of when the day began. While the most elaborate timepieces were shared across regional boundaries in Eurasia, more practical devices – like reliance on bells – varied considerably as well. Common trends and interests there were, but they could be translated in many different ways.

Note

1 The spread of the Gregorian calendar was often a bit painful, even within Western Europe. Britain held back until 1752, keeping the Julian version because of distaste for accepting an innovation from the Catholic Church. But it did finally give in because of the desirability, amid growing trade and contact, of coordinating with its neighbors – a motive that would later become global. But the British move did occasion some debate, with conservative politicians lamenting the shift, bringing the topic up in political campaigning in 1754. However, a widespread story, that ordinary people rioted against the transition because they "lost" 11 days – the adjustment required a ruling that September 2, 1752 would be followed by September 14 – makes for good reading but is not true. The fiction held that rioters chanted "give us back our 11 days" and even worried that their lives would be reduced by those days – again, a great yarn but almost certainly false. More probable is another story, that a clever man bet his friends that he could dance nonstop for 12 days and 12 nights, and then proved it by dancing from the night of September 2 to the morning of the new September 14, collecting some money in the process.

Further Reading

(See also the suggestions in the previous chapter)

On calendars:

Asiapac Editorial. *Origins of Chinese Science & Technology*. (Singapore: Asiapac Books Pte Ltd, n.d.).

Eastwood, Bruce S. *The Revival of Planetary Astronomy in Carolingian and Post-Carolingian Europe*. (London; New York: Routledge, 2017).

Nothaft, C Philipp. *Scandalous Error: Calendar Reform and Calendrical Astronomy in Medieval Europe*. (Oxford, New York: Oxford University Press, 2018).

Richmond, Broughton. *Time Measurement and Calendar Construction*. (Leiden: Brill Archive, 1956).

On the hour:

Cipolla, Carlo M. *Clocks and Culture: 1300–1700*. Reissue edition. (New York: W. W. Norton & Company, 2003).

Dohrn van Rossum, Gerhard. *History of the Hour: Clocks and Modern Temporal Orders*. Translated by Thomas Dunlap. 2nd ed. (Chicago, IL: University of Chicago Press, 1998). A major contribution in the history of time in this period and beyond.

On the use of bells in timing:

Milham, Willis Isbister. *Time & Timekeepers: Including the History, Construction, Care, and Accuracy of Clocks and Watches*. (London: Macmillan, 1923).

On early clocks:

Lloyd, H. Alan. "Mechanical Timekeepers." In Charles Joseph Singer, E. J. Holmyard, A. R. Hall, and Trevor I. Williams eds., *A History of Technology: From the Renaissance to the Industrial Revolution, c. 1500–c. 1750*. (Oxford: Clarendon Press, 1957).

Needham, Joseph. *Science and Civilisation in China: Volume 4, Physics and Physical Technology, Part 2, Mechanical Engineering*. (Cambridge: Cambridge University Press, 1965).

Needham, Joseph, Wang Ling, and Derek J. de Solla Price. *Heavenly Clockwork: The Great Astronomical Clocks of Medieval China*, 2nd edn. (Cambridge: Cambridge University Press, 1986).

North, John. *God's Clockmaker: Richard of Wallingford and the Invention of Time*. (London; New York: Bloomsbury Academic, 2007).

Turner, Anthony J. *Time Museum Catalogue of the Collection Vol. 1 Part 3*. (Rockford, IL: Time Museum, 1984).

On time and Roman courts:

Chrout, Anton. "Legal Profession in Republican Rome." *Notre Dame Law Review* (1954): 91–148.

On European and merchant time consciousness:

Le Goff, Jacques. "Merchant's Time and Church's Time in the Middle Ages." In *Time, Work and Culture in the Middle Ages*, translated by Arthur Goldhammer. (Chicago, IL: The University of Chicago Press, 1980).

Macey, Samuel L. *Clocks and the Cosmos: Time in Western Life and Thought*. (Hamden, CT: Archon Books, 1980).

Scattergood, John. "Writing the Clock: The Reconstruction of Time in the Middle Ages." *European Review* 11 (2003).

Renouard, Yves. *Les hommes d'affaires italiens du Moyen Age*. (Paris: Armand Colin, 1949).

Werner Sombart emphasized the importance of the monastic contribution to the awareness of time: Sombart, Werner. *Der moderne Kapitalismus. Historisch-systematische Darstellung des gesamteuropäischen Wirtschaftslebens von seinen Anfängen bis zur Gegenwart*. (Berlin: Duncker & Humblot, 1986).

On Buddhism and time:

Bunnag, Anawat. "The Concept of Time in Philosophy." *Kasetsart Journal of Social Sciences* (2017).

On Benedictine time:

Kleymann, Birgit and Hedley Malloch. "The Rule of St. Benedict and Corporate Management." *Journal of Global Responsibility* 1 (2010): 207–24.

Zerubavel, Eviatar. *Hidden Rhythms: Schedules and Calendars in Social Life*. (Berkeley: University of California Press, 1985).

On the origins of the call to prayer in Islam:

Haykal, Muhammad Husayn. *The Life of Muhammad*, 4th ed. (Oak Brook, IL: American Trust Publications, 2005).

4 The Rise of the Clock, 1400–1800

Here are two markers, one at each end of a new period in the history of time, in which precision of timekeeping measurably advanced:

1. Clock towers were constructed, often at great expense, in many European cities from the 15th century onward. They could be major sources of urban pride as well as conveying reasonably accurate time to the urban population in a period when private access to timepieces was still quite limited. Their sponsorship by municipalities also signaled a partial move away from the centrality of religion in defining time.

The astronomical clock in Prague (Czech Republic), built in 1410 and still going strong, is a classic example – and a major tourist attraction today. The clock was mounted on a wall of the town hall. It features the movement of sun, moon, and key stars; statues of several major Catholic saints pose on each side of the clock; and it tells time. At each hour, figures of the apostles and other religious figures, plus a skeleton representing death, strike the time on a bell, while a calendar dial represents the months. Local legend holds that the city will suffer if the clock is neglected. The clock was first built by a clockmaker and a university professor of mathematics and astronomy, working in tandem. It was frequently repaired in later centuries, including after the damage inflicted by Nazi forces in World War II.

The Prague clock is an unusually early example of a new urban commitment to time, and somewhat more religious than slightly later clocks that tended to feature more secular figures striking the time – or, as we will see, simply concentrated on telling the time. But its cost, and its central position in the city, were characteristic, signaling the opportunities created by the increasing sophistication of mechanical clocks.

2. Benjamin Franklin, the colonial American printer, in 1746:

> Remember that time is money. He that can earn ten shillings a day by his labour, and goes abroad, or sits idle one half of that day, though he spends but sixpence during his diversion or idleness, it ought not to be reckoned the only expense: he hath really spent or thrown away five shillings besides.

Linkage between time and commerce was clearly heating up by the 18th century, going beyond the use of clocks and into the assessments of character.

The five centuries after about 1300 were unusually significant in the history of time as a whole. Two developments predominated. First, mechanical clocks began to be widely developed and deployed, with steady improvements in technology; innovations soon extended to the first introduction of watches and, at least as prototypes, alarm clocks. The second change involved a growing sense of time and an expanding use of time – there may have been more discussion of time, relative to other topics, in 18th-century Europe than ever before or since. Here, obviously, was where the spread of clocks increasingly intersected with new human experience. This chapter centers on these twin developments.

The centuries in question are often defined, in both European and world history, as "early modern" (though the start date, in 1300, is a bit on the early side compared to conventional definitions of the period). The rise of clocks and changes in the sense of time highlight the importance of the early modern period as a time of key transitions. The same developments also touch base with several more general world-historical questions about these centuries, and may help provide some answers.

For, beyond describing the key changes, the chapter also grapples with a number of analytical issues, most of them interlinked. First, as in any discussion of major change, there are questions both of magnitude and of quality. How widely were the changes experienced? Was this a revolution in time-consciousness or a more limited or selective shift? This is an unusually complex issue for this period, because many very talented scholars have overshot the extent and particularly the speed of change. We know that, by the 18th century, new time-consciousness was gaining ground and that, even earlier, the technology of timekeeping began to improve steadily and significantly. But this does not mean that a revolution in the human interaction with time occurred early on, or that it was full-blown even by 1800. Real change did occur, but the process was more convoluted, and possibly more interesting, than might be imagined – for a modern time sense is not easily acquired.

And on the evaluation side: was the development that did occur good, bad, or indifferent? Did people's lives improve as a result of the changes this chapter explores, or is this not a relevant or answerable question?

Issue number two brings in geography and comparison, around relationships to the West in a period of complex global interactions. The key changes that did occur centered on Western Europe. World historians, particularly in places like the United States, are usually – and properly – at pains not to exaggerate the Western experience in attempting to deal with global patterns overall. But in this case, the West really did lead the charge (again, for better or for worse). And one of the most extensive studies of the history of time, by David Landes, gleefully emphasizes this Western edge. Obviously, world history encompasses many other important cases where a particular society other than the West took a lead – China had led the world in technological innovation from the classical period onward, Arabs long predominated in extending interregional trade. It is not unusual in world history, in other words, to recognize that one particular society pioneered in a particular category of innovation. But given a long tradition of overemphasizing the West, undue Western focus creates some nervousness in a world history context, and the need for sensitive comparative balance has to be recognized. We have encountered elements of this comparative challenge in the previous chapter, and it becomes more important in the centuries when clock time gained ground.

Certainly, a Western lead raises the question of how other societies responded, and this chapter will deal with that issue, for this first age of clocks, directly. But the more sweeping questions are, first, whether Western clock leadership reflects some special Western quality; and, second, whether early leadership in modern timing led to other Western advantages in this period and beyond.

Of course, issues #1 and #2 are related: if even the Western change was gradual and complex, the relationship with the rest of the world may not have been as invidious as the case seems at first glance. Western "leadership" may have been harder to discern than we imagine.

Finally, and this relates to the Western advantage angle, there are questions of causation and impact. The whole emergence of the clock and an arguably new sense of daily time occurred well before the industrial revolution and the other shifts that clearly set modern society apart from its premodern, agricultural predecessor. Did a modern time revolution predate other key developments and, if so, why – why the push for a new sense of time in a still-agricultural context, albeit an increasingly commercial one? And was this shift in time sense fundamental in leading to the more familiar changes – steam engines, factories, huge industrial cities – that would come a bit later on?

The first age of the clock focuses on several major innovations. It also challenges world-historical analysis on a number of fronts.

More Clocks, Better Clocks: The Steady Progress

There are two trends to track in dealing with the major changes in clocks and clock availability in early modern Europe. The first involves a fuller sense of the great enthusiasm for installing public clocks, once the advent of the mechanical clock made this feasible. Public clocks were, without question, the most important innovation in the daily time experience of most urbanites, at least until the 18th century. The second trend highlights the steady improvement in the technology and variety of clocks, where the artistry and ingenuity are clear but the immediate significance somewhat less so.

Municipal Competition

Clock enthusiasm in the cities built quickly, particularly in the 14th and 15th centuries and, while some of them had the fancy apparatus featured on the Prague city hall, others were designed simply to tell the time. A town clock in Milan, Italy, in 1336, struck on the hour, with numbered bell strikes to differentiate the hours themselves; as one observer noted, "this is exceedingly new for people of all estates". By the middle of the 14th century the bells of the clock tower in Florence could be heard all over the city. Strasburg, now in France, had a public church clock by 1354, and then a municipal clock tower by 1372. The French king Charles V, also in the 14th century, built a public clock on the royal palace in central Paris, and sponsored several other clocks in different parts of the city. His palace clock took eight years to construct; the weight that activated it topped 500 pounds. But its single hand only told the hour and, by the end of the day, it could be as much as two hours off. The king of Spain established a public clock in Barcelona in 1392, to allow citizens to lead orderly lives, as he put it, and to "call sleepers and idlers to virtuous works".

The frenzy of clock tower building reflected both a quest for prestige and practical needs. A small town in Silesia, in east-central Europe, petitioned its ruler for "a clock equal to the one in Breslau or better" – Breslau was the nearest big city. Not only having a tower, but being able to brag about the number of bells it contained, became a big deal in urban rivalries. In some regions – Switzerland, for example – even small towns and churches had clocks. A few rich villages also joined in. Italy was less systematic, and some travelers complained about towns where it was hard to find out the time. Scandinavia lagged altogether, joining the clock craze only a bit later. Overall, however, it has been estimated that 500 new public clocks were built in Europe between 1300 and 1450.

These clocks were typically quite expensive. Bells had to be purchased that often weighed several thousand pounds (though some town clocks dispensed with bells and just featured a clock face). Furthermore, someone had to be employed to keep the clock wound and running – another cost. Lots of clocks did not function well, and had to be repeatedly repaired. Not surprisingly some urban residents, including merchants who might be assumed to be particularly enthusiastic about time, complained about the cost.

On the other hand, support for the urban effort came from many quarters, including of course those magistrates who wanted to assure their town's prestige in the rivalry game. Church leaders and monks often urged clocks, sometimes more eagerly than merchants – after all, the Christian commitment to time went back a long way. Regional princes, or kings as in France and Spain, might provide the impetus. Charles V wanted clock time for the efficient operation of his bureaucracy and for more orderly urban activity in general – keeping time became a symbol for well-organized political life. Charles even ordered the churches of Paris to coordinate their bells by agreed-upon clock time, another example of how control of the time continued to be linked to power. Interestingly, Charles himself, a very orderly man, used timed candles inside his palace to organize his periods of prayer and of work – as we will see, it is important not to overdo the clock at this point.

But the change, over a relatively few decades, was real. Mechanical clocks replaced water clocks and sundials in most European cities; clock-timed bells conveyed the time to large numbers of urban residents; and while religion was still involved, the impetus for knowing and controlling time became increasingly secular. The clock itself, with its mechanical intricacies, interested many people, from artisans to intellectuals. Many urbanities would surely have agreed with the French writer Froissart, who wrote a love poem to clocks in 1369:

> The clock is, when you think about it,
> A very beautiful and remarkable instrument,
> And it's also pleasant and useful
> Because night and day it tells us the hours
> By the subtlety of its mechanism

Even when there is no sun
Hence all the more reason to prize one's machine
Because other instruments can't do this
However artfully and precisely they may be made
Hence we hold him for valiant and wise
Who first invented this device
And with his knowledge undertook and made
A thing so noble and of such great price.

Technological Innovation

Enthusiasm for public clocks, from various quarters and for various reasons, also provided clear impetus for often dramatic efforts to improve the mechanisms involved. For mechanical clocks not only required a lot of upkeep, they were often inaccurate. Finally, while public clocks provided great service, some people might well be interested in other kinds of clocks that could be located more privately and that could be consulted for the time on a more regular and personal basis.

In fact, the steady improvements in technology from the 14th century onward are even easier to chart than the spread of public mechanisms, and they were both impressive and important. The clockmaking craft progressively expanded in the European cities, and artisanal inventiveness was encouraged. At the same time, details of the new technology, or at least its packaging, raise interesting issues of impact and significance. More sophisticated technology does not always have predictable consequences.

Initial focus, understandably, involved trying to improve accuracy. The English writer Chaucer correctly lamented, in the 14th century, that "full surer was the cock's crowing as in a clock or any abbey horloge". But in 1500, a major improvement was devised by Peter Henlein, or Hele, in Nuremberg: using a spring rather than a weight to activate the mechanism. Springs were potentially more reliable than weights, and they required less frequent winding; ultimately, they also allowed the construction of smaller timepieces. These gains were not realized immediately: initial springs lost their tautness as a clock wound down, making the later hours of operation longer – that is, less accurate – than the ones recorded when the spring was wound tight. Even so, an early spring clock in England – which could run for up to 40 hours – prompted great admiration: "Every day finer things are being invented – they excite the admiration of the most learned mathematicians."

With springs instead of weights, smaller, even portable clocks could be crafted. Already Charles V had commissioned some smaller household clocks – "And he made his clocks strike in his halls and in his chamber." But this was unusual, and we have seen that the king continued to prefer marked candles when he really wanted precise time. But by the 16th century, with springs, household clocks became more common, some with bells to note the hours. Springs themselves were recurrently improved, making their uncoiling more regular.

By 1500 this development also permitted the manufacture of the first watches,[1] initially in small tubes that were portable though not very convenient. Here is a description of early efforts in making what were first called "clock watches", in Nuremberg, in 1511:

> Peter Hele, still a young man, fashions works which even the most learned mathematicians admire. He shapes many wheeled clocks out of small bits of iron, which run and chime the hours without weights for forty hours, whether carried at the breast or in a handbag.

In the 16th century also, some watches were made for women, called "arm watches" because of where they fastened with straps. But these watches were extremely inaccurate – by several hours in a given day. And the watch faces only showed the hours in any event. They were undoubtedly worn as jewelry and novelties by the fashionable members of the nobility. But use of a better mainspring allowed some notable improvements several decades later. By the mid- 17th century many upper-class men were carrying what became known as a pocket watch, and a fashion of wearing vests, with the necessary pocket, became more common, spurred in England by British royalty.

Early interest in creating miniature devices included some revealing fancies, particularly on the part of royalty. Francis I of France had a timepiece put in the hilt of a dagger, for which he paid a small fortune. Elizabeth I, in England, had a watch put into a ring, which not only told time but served as an alarm, pricking her finger at a designated moment. But most of these devices were mainly for show, not really used for time-telling. On a more practical basis, the capacity to produce smaller mechanisms generated a wide variety of table-top clocks, as well as watches. One historian, David Landes, has noted that the arrival of private clocks and watches freed some people from the dominion of government- or church-controlled public time; they had more personal ways to schedule and measure.[2]

These new devices facilitated one other huge gain in timekeeping: the possibility of activating a minute hand, as well as a hand that would designate the hours. Minute hands had been rare before 1670, but now they became common.

Innovations continued in the 18th century, though there were fewer major gains in clock mechanisms directly (as opposed to time-based navigation devices, to be discussed in a later section). One improvement involved hollowing out the pendulums of big clocks and filling them with liquid mercury, which helped counter the effects of temperature changes on the metal device itself, promoting precision regardless of season or weather. Operating times increased: by the early 18th century some clocks were devised that could run for several days. In watches, by the early 18th century the use of jewels in the mechanism also improved accuracy. A great deal of work went into improving gears and oils, to reduce friction and prevent dust. In London, early in the 18th century, a mathematician from Geneva and two refugee French watchmakers focused on the use of jewels (sapphires) in key bearings to resolve some of these issues. The best watches, by the 18th century, were at most one or two minutes off, per day – a huge gain, more than adequate for most purposes.

It was in the 18th century, also, that experimentation with the faces of watches and small clocks sought to develop the best way to show minutes and well as hours.

Some dials continued to reflect separate rotaries for day and for night – and many of these were quite colorful. But by the middle of the century the best craftsmen were scorning these complexities, and the single dial with two hands (minute and hour) became common. Even then, a persistent custom urged use of Roman numerals for the hours, Arabic for the minutes. But in a further step toward precision from the 1690s onward, in some watches a dial for second hands was added, though it was a young Swiss watchmaker who perfected this addition, in 1776. These watches were sometimes called "doctor's watches", because of a presumed use in medicine, but they had other functions as well – including helping the user to be sure that the watch was running.

Technological innovation is not, however, the only story line in the development of clocks and watches in the 17th and 18th century: elaborate artistic decoration was the other leading concern.

Thus table clocks were typically extremely ornate. Many were shaped like lanterns, with scroll work and other decoration. Some clocks provided calendar and even astrological information on their face, and not just daily time – these functions required adjustment to take into account the differing lengths of the months

In Britain and elsewhere, the grandfather clock emerged – a fairly large clock face atop a cabinet made of fine would, within which the pendulum could operate. Comparable French floor clocks were more elaborately shaped, with more curvature. Table clocks were also ornate, with a variety of decorative themes. Other novelties continued to be developed: the cuckoo clock, for example, was invented in 1775.

Similar trends applied to watches, even as they became smaller. Gold or jeweled casings were common, though elaborate enamel covers were particularly popular. Many cases included designs of the stars or flowers; one model was embossed with a skull, known as the death's head watch. While men typically put watches in their pockets, now a standard feature in vests, women either pinned jeweled watches to their clothing, or carried them on a small chain in a device known as a chaletain.

All of these developments, technological and decorative alike, reflected a number of different interests and contributions. The involvement of scientists was crucial at many points. Huygens, the inventor of the pendulum, became involved in some angry disputes about whether the pendulum's movement changed with the weather. Some claimed that pendulums moved faster in warmer seasons or climate, but Huygens stubbornly insisted (incorrectly) that there was no temperature effect. The link with wider scientific studies of gravity and motion is obvious as well, particularly by the 17th century – as Galileo's involvement suggests.

But there were connections also to huge advances in furniture and jewelry design. This was a great age of furniture making in Europe, as the prosperity of some households increased. Elaborate cabinets and finely crafted chairs were increasingly popular. Clocks, certainly, became part of this artistic trend, making it difficult at some points to determine whether they were sought because they kept time or because they adorned a salon.

Many developments, and certainly the spread of manufacturing, linked to the emergence of a solid and imaginative craft of clock- and watchmakers, with centers in many parts of Europe – Nuremberg in Germany, Switzerland, and several locations in Britain, France, and the Netherlands. Many artisans moved from place to place, easily crossing national borders. Some early specialists entered this field from the older craft of locksmithing, and other groups, like goldsmiths, could also get into the act because of the importance of luxury fittings. But by the 16th and 17th centuries a powerful separate artisanal specialty had been established. True to the age, clockmakers in many areas applied for, and received, guild status, in some cases chartered by the king. But while many guilds were best known for their attachment to existing techniques, many clockmakers continued their commitment to technical innovations, undoubtedly providing the most important single source of recurrent technical change. Not surprisingly, many laymen, both scholars and upper-class, professed great admiration for what clockmaking involved, sometimes dabbling themselves and frequently using clock mechanics as images of precision and regularity.

The advances in clockmaking and the resulting diversity of products were truly important, but they raise some obvious interpretive problems – including the recognition that, at least until the later 18th century, many of the items did not provide great accuracy.

Most obviously, most of the clocks and watches were quite expensive. Master clockmakers had to work slowly, and their production was accordingly limited. One shop, in the core city of Nuremberg, generated 6,000 watches in a lifetime. This was a prodigious output in one sense, but it could supply only a minority (the city itself had a population of 30,000, and many of the clocks and watches were exported to other areas). Limited supply and substantial cost confined access to individual timepieces to an upper class, with a few exceptions to be noted later. An interesting division opened between the small number of people with clocks in the home or watches in the vest, and the bulk even of the urban population. A gender disparity may have operated as well, as aristocratic men were far more likely to sport watches than their female counterparts. Of course, the majority of urbanites usually had access to time through the public clock towers and their bells, but personal recourse, and such niceties as dials that noted minutes, or even seconds, as well as hours, was limited.

The increasing professionalization of clockmaking, including the formation of guilds, tended also to support the emphasis on very expensive production, rather than experiments toward lowering costs and generating wider use – at least until the very late 18th century. Guilds typically tried to discourage competition, and clockmaking turned out to be no exception despite the tremendous contributions individual artisans had made, and continued to make, in technical development. Production of artistic products for an elite seemed eminently satisfactory to most relevant craftsmen, who could also boost their income by providing the regular servicing that many of the home items required.

And this leads to the second analytical challenge, at least as interesting. It is not at all clear what the main motivation was for most of the people who could buy table clocks or watches. The 17th and 18th centuries in Western Europe constituted what many have seen as a first age of consumerism. People became more interested than before in acquiring fashionable, colorful clothing and more elaborate home furnishings, and while historians have been particularly interested in charting the rise of popular interests, the upper classes displayed consumerist appetites of their own.

And clocks and watches easily fit the bill, serving as early examples of what would later be called conspicuous consumption. Hence the tremendous, and technically unnecessary, concern for fancy trappings, expensive materials and designs, and even, at least initially, a number of items, like the early watches, that were not particularly useful in telling the time in any event.

Putting the case simply: it is not easy to be sure whether clocks and watches (again, aside from the public installations, though some of these had their own luxury flourishes) were being consumed because they helped tell the time or because they constituted novelty, impress-your-neighbor, additions to furniture and personal adornment. Indeed, there is considerable evidence that, even in the 18th century, a number of men were buying pocket watches with no interest in the time, but simply as a means of personal display.

Again, there is uncertainty. Britain was producing at most 200,000 watches a year by 1800, many of them for export. At the price of three to five pounds, only a prosperous master artisan would have been able to afford this item. But it was estimated, in 1800, that 700,000 households in the nation had a watch or clock, and at 20 shillings many people might have afforded at least a wooden clock. Some workers viewed a clock as an investment, to be purchased when there was a run of luck, then frequently pawned when money was tight. And there is no doubt that purchases were rising by the final decade of the 18th century. Ambitious research efforts have not yielded definite conclusions about the extent of interaction between ordinary people and personal timing devices.

Unquestionably, the spread of timepieces from the 15th to the 18th century attracted the interest of many scholars and intellectuals at the time, particularly in the areas of mathematics and astronomy. It reflected royal interests in good order and claims to power. It caught the fancy of many observers, convinced the better technology in this domain was a sign of progress. And, gradually, it did both encourage and permit people to develop a more precise sense of time and to apply this sense in various facets of daily life. But this development was gradual and, even by the 18th century, far from complete, even in the urban upper classes. And by definition it left out the majority of the population, still rural: a few villagers near big cities might occasionally catch the sound of bells, if the wind was right, and they might be exposed to time if they needed to know when urban markets opened or the city gates closed – but many people were not involved even to this extent.

So what, during these long centuries of gestation for the mechanical clock, amid impressive bursts of genuine technological progress, were the uses of time, and how much did these uses change?

Time in Daily Life Before 1800

Judgments about actual uses of time are complicated by several factors. The most important is simply the issue of evidence. A number of historians have been extremely inventive and thorough in ferreting out data about when people began to make use of units like the hours of the day, and not simply the indications provided by nature – and also when they continued to ignore these niceties. But even with best efforts, there is much that is unknown, simply because relatively ordinary people do not always leave much record and because, even when documents exist, they may not refer to something as prosaic as time. It is far easier to pick up the excitement about new clock technology or beautiful designs than to talk with much assurance about actual time sense. We can make use of some really imaginative research, but the limitations are inescapable.

The second dilemma is also common in historical work. We know, unquestionably, that a new sense of time did ultimately develop. There was real change. Fascination with time measurably increased by the 18th century. And by the 19th century, if not before, ordinary people were bent to the rhythm of clocks and watches in work and school and beyond. But knowing the ultimate result is not the same thing as knowing how rapidly or how early fundamental change began to take hold. There is always the risk of exaggerating early evidence of new time use because we know what definitely existed later on.

And, finally, this common dilemma, of trying to develop an accurate picture of often gradual change even in light of a known result, is exacerbated in this case by some arguably excessive scholarly enthusiasm. For, excited about the importance of time in modern human history, some very eminent scholars (often not professional historians, for what that is worth) have issued stirring statements about a revolution in time that may confuse what was possible technologically with what actually took place in the complexities of ordinary life.

Here are three examples. First, from Bilfinger: "The entire evolutionary process of urban life was pushing with a certain natural inevitability, toward the environment of the striking clock." (Just for the record, many historians get a bit nervous around the idea of inevitability, which may encourage neglect both of complexity and causation when assessing change.)

Or Crombie: "The invention…of the mechanical clock, in which the hands translated time into units of space on the dial, completed the replacement of 'organized' time…by the abstract mathematical time of units on a scale belonging to the world of science."

Or philosopher Lewis Mumford, in his 1934 classic on *Technics and Civilization,* where he argued that clocks, not steam engines, were the key force that created the modern world:

> The clouds that could paralyze the sundial, the freezing that could stop the water clock on a winter night, were no longer obstacles to time-keeping: summer or winter, day or night. One was aware of the measured clank of the clock. The instrument presently spread outside the monastery; and the regular striking of the bells brought a new regularity into the life of the workman and the merchant. The bells of the clock tower almost defined urban existence. Time-keeping passed into time-serving and time-accounting and time-rationing. As this took place, Eternity ceased gradually to serve as the measure and focus of human action.

Big claims, but claims that may not do justice to the actual complexities of change and continuity in this vital aspect of human life, claims that may exaggerate the precision of available technology and actual adaptation to this technology. Let us look at some evidence, both for signs of change and – a more difficult category – indications that earlier and more imprecise patterns still prevailed.

The advent of town clocks undoubtedly firmed up the existing use of bells to indicate key passages in the urban day – when markets opened, when a city assembly was to meet, the opening and closing of the town gates. Some morning bells may also have indicated, at least loosely, when work was to begin. With clocks, these traditional patterns could become more precise. The city of Dresden thus set complex schedules for the public markets: only locals could shop before 8, bakers could buy their supplies between 8 and 10, other categories thereafter.

Other evidence, from as early as the 14th century, is even more interesting. In 1395, for example, a German notary, signing off on a document, indicated that the clock had just struck five hours after midday. Notaries were key figures in many parts of Europe, because of their role – as literate individuals – in concluding various kinds of legal documents. The suggestion that they were beginning to incorporate time in their certifications is thus fairly important – though we do not know how widely or quickly the practice spread. And of course the reference to hours reminds us that it was still too early to worry about minutes – which available timepieces did not yet indicate.

By the 14th and 15th centuries scattered evidence, particularly from Italy, suggests that parents or, more properly, fathers, in merchant families, were beginning to record the time of the birth of a child or grandchild. One such case pops up in Nuremberg, that clockmaking center, as early as 1375. Leon Battista Alberti, artist and author of a famous and early Italian book on manners in 1434, urged this practice – "it shows the conscientiousness of a father". Presumably, this kind of exactitude might conceivably be important later on, in case of a contest over a parental will – it would be vital to know precisely who was born first.

Also from 15th century Italy, there were occasional references to the hectic pace of merchant life, marked by appointments on the hour. As one man stated, "I don't have any time. It is the twenty-first hour and I have had nothing to eat or drink." We have seen that this kind of lament cropped up earlier, in classical Rome, but without quite such precise allusion to clock-based time.

There are also strong indications that many monasteries by this point switched to clocks, even scheduling activities on the quarter- or half hour. This would extend existing reliance on hourly time, but again with greater precision.

Some cities began to regulate times of work, though in very general ways. Cologne in the 14th century stipulated the hour that day laborers should be allowed to eat lunch, and also the end of the work day: 10 PM, for belt-makers. Blacksmiths, on the other hand, were only to labor from 8 to 5. A good bit of work was still scheduled by the position of the sun, and combinations with clock time were probably common, but there was a movement toward the clock. Just for good measure, the Cologne city fathers also insisted that no fish was to be sold before 6 AM. And there were a few more dramatic examples. In one English town a businessman left an endowment such that the parish church bell would be rung, for a full half hour, at four in the morning during winter time, to allow people to get ready for "the labours and duties of their several callilngs".

City courts as well as assemblies began to shift to clock time, rather than relying on church bells that simply indicated one of the hours of worship. Thus a court in Paris in the late 14th century stipulated that litigants should appear at 9 hours by the public clock "or approximately". And, increasingly, fines were imposed for tardiness.

Revealingly, some schools (where however only a minority of urban children attended) began to set starting times. A Jewish school in Germany in the 14th century gave its students "a specific time for coming and going". In Nuremberg by 1515, subjects of study were divided on an hourly basis – math one hour, reading another and so on. And some schools began not only to use clock schedules but to talk more generally about the importance of economical uses of time – a theme that would gain strength later on, as Ben Franklin's aphorisms would suggest. By the late 17th century schools in England were known for their insistence on strict start and stop times, and this pattern could well socialize children to time quite early on. Yet in this vital category, there are some caveats. First, we do not know how widely or quickly these school schedules spread. Second, the schedules often relied more on hour glasses than on clocks – thus in Moravia as late as 1676 a regulation intoned, "no teacher may teach without a sandglass". Further, many children still had no contact with school at all. And finally, intriguingly, universities, notorious for long lectures, did not at this point move to a clock-based schedule. The turn to clock time was real, but uneven.

Not surprisingly, as clock time became more valued, hours began to be set for baptisms, weddings, and funerals, though Protestant regions moved toward this more quickly than Catholic. In an entirely different vein, there is evidence that the British began to time horse races by the late 17th century – though an actual stop watch would only be invented (though also for horse racing, for the king of France) in the 1820s.

By the 18th century, as the British began to build a better road system and merchant travel increased, coach services began to compete for speed. It took two days to go from London to Bristol in 1754, but 30 years later a good driver could do this in 16 hours. Drivers were given watches to hold them to their pace, while passengers, also with watches, checked the time as well. The impact of time competition on horses was not commented upon, but there is no question that the animals suffered under this new regime. Advertisements announced precise times for departure and arrival. Here was preparation for the impact of early railway systems on time, to be discussed in the next chapter.

Also in England, by the late 18th century, postal deliveries were rigorously scheduled (for three times a day) – so people who saw the postman in action could have a pretty good sense of what time it was.

It is also possible that, by the late 17th century and particularly in the 18th century, a number of urban families began to introduce clock time into family meals, or at least dinners. Before this point many people had eaten irregularly. But with new prosperity, including more elaborate table settings with fancier china and utensils, many wives and mothers began taking new interest in regularizing the family meal. This could easily extend to the notion of regular timing, and expectations that family members should show up promptly or face complaints. This aspect of family time, and the idea of criticizing tardiness, extended into the 19th century even as a growing number of family members began working outside the home.

So, overall, there is little question that clock time increasingly mattered, in various aspects of life in the cities. The accumulation of uses was gradual but steady. But, lest we prematurely justify the notion that a modern time sense had been born, some complications must be evaluated as well.

First, many people continued to prefer the old day–night distinction, by which the length of hours would vary with the season. One formulation in 1487 noted that 52 days of the year had eight hours, 17 had 13, 22 had 15, and so on. On this basis clocks could be adjusted seasonally to shorten or lengthen the hour as needed. Also, individual cities continued to maintain distinctive definitions of the day – at what hour noon occurred, for example. Travelers often took cheat sheets with them, so they could keep up with the particular clock schedule operative in each town they visited. Tradition and variety were hard to shake off, even when clocks came into greater use.

Second, it is also vital to remember how inexact many clocks and watches were, which could limit reliance on hourly scheduling and which certainly required continued use of older devices. Many people carried pocket sundials with them as late as the 18th century. Where it was important to set a time limit on testimony or speeches, sand glasses were more commonly used than watches. All this reflected time needs, to be sure, but not of an entirely novel sort. Add to this the extent to which, for many people, watches and domestic clocks were valued for their show value more than their role in timing, and the complexity is confirmed. Public clocks may well have remained more important than more personal devices, but this in turn imposed some limits on uses of time in indoor settings.

Third, and most important, a number of vital human activities were not yet timed very clearly at all, despite the fact that, technically, they could have been. Five examples can be cited, though these may only scratch the surface in an age in which time sense was, at most, in transition, though not yet revolutionized.

1. Scientists used clock time less than one might imagine. We have seen that mathematicians and astronomers were very interested in time and clocks; the use of clocks to make astronomical observations more accurate was really important. But other branches of science largely stood apart. Even as researchers became more interested in the circulation of the blood, and began to talk more about the pulse, they did not connect their work to timing – beats per minute – until at least the 18th century, despite the existence of "doctor clocks" with minute and second hands, which would have allowed them to do so. Thinking in terms of using the clock was not automatic.
2. Much work, most work, remained untimed. Many people continued to alternate between periods of intense work and slacking off, with no clock timing involved. The biggest change in European labor through the mid-18th century was the rapid expansion of what was called the domestic or putting-out system. Here, people worked in their homes (urban or more commonly rural), using simple tools like spinning wheels or hand-operated looms. They sold their products to urban merchants or actually took orders from these merchants. But while merchants might press for early completion, in periods of strong sales, there was no clock involved. Workers got started and finished at their own pace,

sometimes mixing in some agricultural labor. To be sure, uses of town clocks to measure the work day for some artisans constituted significant developments, but clock-regulated work was not yet the predominant pattern.

3. European militaries were slower to turn to clock time than one might imagine, though ultimately some change emerged. In principle, using clocks to schedule military maneuvers might seem a no brainer: using banners, voices, drums, and horns to get a troop going only works up to about 600 men – the sounds will not carry farther. It would have been a real advantage to arrange a time at which larger numbers could be coordinated. But clocks were often too inaccurate still to take this risk. Shakespeare's plays make several references to aides who were late in waking up the commander, or how it was only the rooster's crow that made people aware that it was time to put on their equipment. Though there are some references to clock time – one troop got going "upon the stroke of four", most armies still were regulated by dawn and dusk. An Italian officer in the late 15th century lamented that his clock had broken down, and asked for a new one "even if it doesn't ring." Some soldiers took their own roosters with them so they would wake up on time (for ordinary soldiers could not afford a watch of any kind). When battles occurred, clocks were not involved: commanders relied on visual and auditory signals to try to move their forces. To be sure, by around 1700 some officers took a clockmaker with them, to repair devices as needed. And by the end of the 18th century officers routinely carried watches. Even as late as World War II the US government provided watches only for combat officers, pilots and navigators – ordinary troops did not necessarily need them. The merger of clock time and military operations was surprisingly complex – and gradual.
4. Leisure was not, for the most part, significantly affected by any new sense of time. There is no reason to believe that young children's play had any particular time component – children were drawn into a time network only to the extent that, a. they went to school, and b. their school had begun to convert to hourly schedules. For most adults, leisure was still largely defined by the festival cycle (though, for some, Protestantism reduced the number of festivals significantly). And festivals were not timed.

   Even newer leisure forms were not clearly time-dependent. In the late 16th and early 17th centuries, the plays of Shakespeare and other dramatists were performed in London theaters, like the famous Globe. Performances occurred in the afternoon – there had to be light – usually beginning about 2 or 3 and running for roughly three hours. But the schedules were not precisely timed. When a play was on the docket, a pennant would be displayed at the front of the theater, and it would be lowered when the play began. More important, trumpets would sound shortly before the performance, to let people know it was time to get their seats. Many people milled around in front of the theater before the show in any event, for there were shops and street entertainments – so no clock-based precision was necessary. (Even today, of course, custom allows a few minutes' difference between official and actual start time in the theater, a remnant of the more traditional approximation.) Trumpet signal, not clock time, was the crucial indicator back in the day.

5. Sleep was not timed for the most part, and we increasingly realize that, in Europe and some other traditional societies, sleep patterns were very different from what they have become today. Many preindustrial cultures, including Europe at least from the 14th to the 19th century, feature two periods of sleep – referred to, not surprisingly, as first sleep and second sleep. In this segmented sleep pattern, people go to bed around 9 PM (though no set time is stipulated, and it will vary seasonally) and sleep for 3–4 hours. Then they will wake up (again, no clock involved, just body habit) and read, pray, do a few chores, chat with neighbors or have sex for an hour or two, then go back to sleep (a lighter sleep, however, than the experience of the first sleep) until morning. Their total sleep time is about the same as what is recommended today, but its pattern is both more complicated and more casual. Some people did depend on some prompt to wake up from the second sleep – cock's crow most commonly – but the possibility of using a clock with a bell chime did increase by the 18th century. (And one American craftsman invented an alarm clock in the 1770s, but only for his own use since he had to get up at 4.) Overall, there was neither need nor experience of talking about sleep in terms of hours or precise time allocation: it just happened. There is, to be sure, debate about whether this pattern is "natural" – anthropologists have identified it in several places, apart from the historical finding concerning Europe. For our purposes, the dissociation from timing is the main point – plus the fact that this relationship would begin to change in the 19th century, when time became more important.

Overall, it seems clear that the uses and sense of time advanced less rapidly, and more haphazardly, than the timing technology did between 1300 and the later 18th century. The proposition is complicated by the fact that timing devices themselves, though advancing, were so often inaccurate –which helps explain the slowness and unevenness in the changes in the human experience. Urbanites, obviously, experienced more pressure than the rural majority, adults more than children. Protestants may have become more interested in timing and timepieces than Catholics were – in some mixed-religion areas, Protestants vastly outnumbered Catholics among clockmakers. But the main point is the complexity of the changes that did begin to take shape, and the intricate balance between timed and untimed experiences in urban life.

There is, however, one final point. From the late 17th century onward, in places like Britain, there were enough time-conscious people around, both moralists (particularly Protestant clergy, convinced of the evil of idleness and the importance of time-consciousness to compensate) and businessmen, to generate a growing concern about the bad time habits of most ordinary workers. Thus, one mill owner, convinced of the "sloth and negligence of those paid by the day", tried to set up a more regular work pattern, marked by bells both for start and stop times and for meal breaks. Some systems of this sort remained a bit tentative: the great pottery maker, Josiah Wedgwood, urged a time-based schedule but insisted only that latecomers "should be noticed" and, perhaps, given a talking-to "after repeated marks of disapproval". The basis for an attack on popular time habits was building, but it had not yet uniformly arrived.

European Clocks and World History

The rise of mechanical clocks in Europe, and their impact (if somewhat limited) on the European sense of time raise vital questions about impacts on other societies. These questions, in turn, intrude on two familiar issues in world history during the early modern centuries: the expanding geography of "the West" and the evaluation of societies that deliberately remained outside the Western orbit. The history of clocks offers a good opportunity to address both these issues with some specifics in mind.

Western Europe was obviously the center of Western civilization during the early modern centuries, but it participated in growing outreach through aggressive trade and colonization. Two or three regions may have been so deeply affected by European influence that they were drawn into a common Western orbit. The Americas were the most obvious case in point, and particularly the British colonies of North America, where a variety of shared patterns jostled against distinctive features of frontier conditions. But Russia is another interesting question mark, particularly after Peter the Great's reforms, around 1700, deliberately implanted a number of Western models. Assessing the results of European clocks on Russia and North America in no sense definitively answers the overall question about Western-ness, but it does provide an interesting touchstone.

The relation of Europe to other parts of the world in this period – where there is no question of any significant "Westernization" – has been seriously recast in recent decades, particularly with regard to key societies in Asia. Some formulations used to assume that, as the West became more dynamic and amid developments like the scientific revolution, the rest of the world simply progressively fell behind. We know now that this is a truly misleading comparative formulation. The Chinese economy, for example, remained extremely vibrant, as indicated by the fact that China, not any European nation, ended up importing the largest quantity of American silver in return for its high-demand products, between 1500 and 1800. Where does the history of clocks and timing fit within this now complicated framework? The question is hard to answer but also hard to avoid.

The context is clear enough, even though the regional variety of responses turns out to be considerable. Europeans had a clear lead in mechanical clocks. As we have seen, knowledgeable Europeans were immensely proud of the ingenuity of their clocks and of their artistic quality. Not surprisingly, clocks were one of the items Europeans took with them on commercial and exploratory trips – often offering them as gifts to local leaders. They were also eager to sell for export. Even aside from gifts and marketing, European ships increasingly used clocks directly, though the clearest advances in navigational clocks did not occur until the 18th century. Clocks and European outreach were an important combination, calling for some reaction from many of the societies with which they came into contact.

Russia and the Americas

Some regions interacted with European clocks fairly readily, at least in some respects. Russia, which would have a complicated history of dealings with the West from the 15th century onward, took to clocks early on. There were two obvious spurs. First,

Orthodox Christians shared some of the same interests in timing daily rituals that Catholics did, though with different specifics. Good Christians were supposed to pray at certain times in the day. And while Russian monasticism did not follow the Benedictine rule, timing helped regulate daily activities in this case as well. Further, Russian Christianity early on began to deploy bells, which sounded on various occasions, among them noting daily religious services.

Second, by the 15th century Russian leaders were becoming actively interested in contacts with the West, particularly in the artistic domain – as in the use of Italian architects to design buildings in Moscow's Kremlin. As Russia gained independence from the Mongols, and given the collapse of their traditional exchange partner, the Byzantine Empire, openness to selective imports from Western Europe increased.

This was the context in which mechanical clocks first appeared in Russia, at an impressively early date. A Moscow prince commissioned a mechanical clock for his palace behind the cathedral, and it was installed in 1404. The builder was an orthodox monk from Serbia (known as Lazar the Serb), whose monastery was also a center of learning in fields like mathematics – the combination of religion and science was ideal. Lazar's ability also reflected the fact that knowledge of clock-making was spreading rapidly in the Balkans prior to the Ottoman conquest of the region. The Moscow clock, installed in 1404, had a face with numbers written in church Slavonic, and a striking mechanism. For a long time, however, this was the only public clock in Russia, though it worked without failure for more than two centuries, and continued to be regarded as something of a marvel.

A few more clocks were built in Russia during the later 16th century, including installations on three of the towers of the Kremlin complex (one of them constructed by a collaboration between a Scottish expert and local clockmakers); one of these played music on the hour, with a 24-hour dial displaying Cyrillic numbers. But the next big move came as part of Peter the Great's Westernization program at the end of the 17th century. Peter himself, during his famous incognito visit to Holland, explored clockmaking shops as well as the shipyards and manufacturing centers. In 1706 Peter purchased a massive clock from Holland, transported in 30 wagons from Amsterdam to Moscow and installed by the watchmaker Ekim Garnov. It had a 12-hour dial on the face. Later in the 18th century it was adjusted to play a German tune on the hour. The great Kremlin clock would undergo many repairs and adjustments, but survives to the present day.

These public clocks were supplemented, particularly during the 18th century, by significant imports of table clocks and probably watches from centers in Western Europe. In broad outline, developments in Russia might seem to mirror rather closely what was happening in the West – a growing interest in time particularly as indicated by the installation of public clocks, but disproportionate emphasis on expensive artistic models in the hands of the upper class.

There were, however, two differences. One is a familiar basic: even by the later 18th century, Russia was a far more rural society than was true of the West (with at most 5% of the population in cities), and it had a very small merchant class. This means that a lower proportion of the Russian population was in any way exposed to modern clock time, instead relying on indications from nature plus whatever timing was provided by religious services. But, second, what Russian clock use there was continued to depend almost entirely on imports. A small group of clockmakers developed in Russia, primarily for maintenance purposes; a few individual artisans crafted clocks by the later 18th century. But this was not a robust response. A small amount of clock manufacturing would emerge in the 19th century, but the dependence on the West for imports would persist essentially until the advent of the Soviet Union.

North American patterns were different. Obviously clocks reached the British colonies at a later date than was true for Russia. Even when the *Mayflower* arrived, it carried no clock. But interest accelerated greatly from the late 17th century onward. A chart of the relative frequency of references to clocks in American publications shows a rapid surge by that point (though a bit later than was true for publications in Britain), with a peak in the mid-18th century. By that point, imports from Europe (mainly Britain) began to increase, though the sheer expense was a major deterrent. By the 18th century local artisans, particularly in New England, were experimenting with clocks with wooden rather than metal parts, from the sake of economy.

By the middle of the 18th century, towns, again particularly in New England, began to sport public clocks, sometimes on churches, sometimes on municipal facilities. Some of these were sponsored by local clockmakers themselves, to advertise their wares. Thus one Ebensezer Parmele, of Guilford, Connecticut, built a public clock for the local church Meeting House, in 1741. But a clock in Frederick, Maryland, later in the century, was paid for by public fundraising, suggesting growing popular interest in knowing the time. By this point also, public clocks were spreading beyond New England, though they remained less common in the South.

By the later 18th century, a number of American artisans were making clocks, reducing the dependence on exports and to some degree modifying the problem of cost. Some of the craftsmen were local, others immigrants with prior training in Germany or Switzerland. In one striking case, Benjamin Banneker, the son of former slaves, borrowed a watch from an acquaintance, took it apart, ultimately using this as a model to build an impressively accurate clock entirely from carved wooden pieces, and then capitalized on the notoriety of this product to set up his own repair operation. Literally scores of shops existed, particularly in Pennsylvania, Connecticut, and Massachusetts, by 1800, and this proliferation set the stage for more substantial American participation in the history of time in the 19th century. Some of these shops were beginning to produce the American version of the grandfather clock.

But more than clocks were involved in the growing American interest. As in Europe, clockmaking attracted the attention of the elite, because of the sheer inventiveness involved. Thus in 1792–3, Thomas Jefferson designed a clock of his own (later installed in his home in Monticello, where it can still be seen). The clock reflected Jefferson's delight in novelty, but also his Enlightenment-style passion for rational order. The device had two faces: one, facing outward, had a single hand to tell the hour – presumably this degree of accuracy was sufficient for agricultural slave labor. But the face for the interior of the house had a minute hand as well, suggesting more precise schedules. And instead of a bell, the clock sounded a Chinese-style gong. Jefferson would also provide a clock for the new University of Virginia.

Jefferson was not alone among the Founding Fathers. Benjamin Franklin also had a lively interest in clocks (and gongs). And it is Franklin, in his writings, who suggests an even more profound American involvement with time, by the later 18th century – in his many recommendations on the importance of timing one's daily life toward greatest possible efficiency. Many Americans, whether sold on time or not, would be drawn into its orbit before 1800.

We know less about European impact on Latin American time practices than we do about North America. There has been a substantial debate among historians about the extent to which Europeans disrupted the well-established sense of historical time that had developed, particularly in central America. One recent argument contends that native peoples retained the ability to think in terms of deep links between past and present, so that the past could be recaptured through ritual practices; but others contend that Christianity unsettled this older framework. On a more prosaic level, we know that peasants in 17th-century Chile monitored the cooking of an egg by the time it took to say two Ave Marias – which reflects European influence but not clocks. Spanish officials in the political centers undoubtedly used mechanical clocks, but the extent of their influence is not clear.

Asia

Patterns in Asia were quite different, certainly from the American experience and even from the Russian. European clocks were imported, mainly as artistic or novelty items, by some members of the upper classes. They drew some attention. But they had no wide impact, and stimulated little local activity until well after 1800.

Two points were particularly distinctive: first, there were few if any public clocks. This meant that the disparity that existed even in Europe, between general urban access to daily time and what was available to the upper classes, was particularly acute. And, second, considerable investment continued to be devoted to older timing devices, though some of these were intended for public use.

Thus in the great Mughal Empire, public clocks seem to have been unknown until well into the 19th century, when they were imposed by the British. But the Mughals did invest in some very elaborate public sundials.

The Ottoman Empire, similarly, displayed little interest in public clocks. The only exceptions occurred in the Ottoman holdings in the Balkans where, as we have seen, a clockmaking tradition had been established by the 14th century, before Ottoman arrival and in conjunction often with Christian centers. Several public clocks were built by Christian communities, for example in Macedonia, during the early Ottoman period. But in the Middle East, primary attention went to refining existing technologies, as with water clocks, often in association with the ritual needs of Islamic worship.

Again, none of this suggested a lack of interest in time; and private imports of European clocks accelerated as well. A contract as early as 1632 thus specified a delivery from Switzerland that included striking clocks and simple watches; lunar-movement clocks were also very popular, with one order in the late 17th century asking for "as many lunar-movement watches as you can make in the space of one year". Inheritance records from the 18th century show that imported clocks were common items in upper-class households – more common than imported pistols. But a commitment to widespread innovation and a sense that older methods had somehow become inadequate, were largely absent. Even when clocks were purchased, lack of local expertise for repair often meant that they were soon confined to artistic display cases, no longer operative; this happened to several imported clocks in the Ottoman royal palace in Istanbul. And, well into modern times, the Ottomans continued to use Islamic time, dividing the day into two 12-hour segments, beginning the day at sunset, and resetting the timepieces every sunset to make sure they started at 12. The tempo of daily life was not significantly altered from earlier patterns.

Most Ottoman cities in the Middle East continued to place great emphasis on older methods. Sundials were widely deployed, and in many cities "timekeeper houses" provided more precise solar observations, particularly of course to indicate the appropriate hours for daily prayer.

Patterns in Japan were somewhat different, though again there was no public clock movement until the 19th century. Various European missionaries brought clocks to Japan by the later 16th century (the first gift was in 1551), prior to the nation's decision to impose greater isolation. One expedition offered a clock as a gift to a feudal lord in 1591, but he (politely) rejected it, saying "I would have no use for this." Part of the problem might have been simply that the clock face used numerals that were unfamiliar to the Japanese. But the biggest issue was that European time, by this point, was departing from well-established Japanese assumptions – even though Europeans themselves had often shared these assumptions at an earlier date.

For the Japanese were committed to the idea that hours should vary in length depending on the season, thus preserving six-hour units for both night and day (obviously, using six rather than 12 hours meant that the unit was longer than the Western standard, quite apart from being seasonally variable). The Japanese saw absolutely no reason to jettison this standard, which worked quite well for them. But after brief hesitation they did figure out a way to adapt the Western mechanical clock: it simply needed a different face (with Japanese figures rather than Western numerals) and it needed to be adjusted roughly every two weeks to take into account the changes in the length of the hour. Astronomers developed detailed calculations that allowed households to determine the beginning of the day (rather than depending on simple observation), and each household had a calendar to guide the clock adjustments.

By the 17th century an active clockmaking craft had emerged in Japan, based on Western techniques but adapted to the Japanese need for flexibility in the length of hours. (The Japanese also began producing mechanical dolls, using similar mechanisms.) Several manuals offered instruction on clock design by the 18th century. It was claimed that some clocks could run for up to a year on a single winding.

Japanese society was in fact very time-conscious. Diaries, contracts, and court documents from the early modern period often carried a time-stamp. Japanese, at least urban Japanese, frequently consulted the time. But they operated by a different system, used clocks in a different way, and – by European standards at least as of the 18th century – were content with quite inexact reckonings. For adjusting clocks only every two weeks meant some significant disparities in time units as the season continued to change. And it is also important to note that the Japanese also used various systems of timekeeping, depending on purpose: one system for religious rituals, one for astrology, another for daily organization. But the overall system had the merit of preserving tradition, and the Japanese made it work.

China poses the biggest issues in assessing responses to Western mechanical clocks, and in trying to figure out the significance of cultural difference. As we have seen, the Chinese had a long history of studying time, seeking greater precision, experimenting with a variety of devices. They had built some of the most elaborate water clocks ever. And they had the most impressive overall history of inventiveness of any society. On the other hand, there is evidence that relevant Chinese artisans sometimes found it difficult to pass knowledge on to others, because of constraints in communication, and some devices were not replicated in consequence. And the imperial court, claiming great power, periodically intervened in timing discussions, particularly insisting on producing new calendars, sometimes arbitrarily, and demanding compliance.

There is no question about how the Chinese began to learn about Western mechanical clocks. The clocks were introduced, in the late 16th century, by the Jesuit missionary Matteo Ricci. Ricci was eager to stimulate an interest in devices like clocks as a means of winning imperial favor, ultimately in order to promote Christianity more directly. And he apparently had his own belief that the superiority of Western clocks was a direct indication of the superiority of Western religion, though he tried to be tactful.

The result of the missionary–emperor, clock–Chinese interactions has occasioned a huge scholarly debate, with some contemporary overtones but also possibly more heat than light. At an extreme, some Western historians have argued that the Chinese refused to regard clocks as anything more than toys, because to do otherwise would jeopardize their intense sense of Chinese superiority. In rebuttal, partisans of Chinese initiative make three overlapping claims: first, that the Chinese court was really interested in clocks, not belittling; second, that (like many societies) the Chinese had some good reasons to be fairly content with existing systems, particularly given the limitations of clocks; and third (most recently), that a clockmaking response did emerge in some Chinese cities, somewhat independent of imperial whim.

Here are some of the components of the debate. Ricci not only brought clocks but also evidence that European calendars were more accurate, astronomically, than the Chinese. When he did win audience with the Ming emperor, the latter had no interest in religion but was absolutely fascinated with the clocks, asking at one point why they stopped chiming after the hour had sounded. Ricci, not a modest man, claimed that his imperial host was "dumb with astonishment", and then kept a clock before him constantly because of his fascination with the hourly bells. Jesuits replaced previous Muslim astrologers in advising the emperor. And later emperors ordered hundreds of imported clocks, among other things giving them as gifts to their children. It is important to remember that clocks at this point were not necessarily more accurate than the timing devices the Chinese already had; their main advantage was that they were more portable.

The pro-West historiography goes on to argue that while Chinese interest was real, it operated at the level of novelty items, not serious respect for the opportunity of keeping better time. It was true – and recent research has expanded on this point – that a number of Chinese artisans began to emulate clocks and set up clockmaking establishments – Chinese reactions did not just depend on emperors. And the ruler himself set up a clockmaking enterprise in the palace as well. But the artisanal efforts may not have been technically adequate. Certainly, by the 17th century, European visitors to China were unimpressed: a Dutch group wrote, "although ingenious and subtle, the Chinese have almost no instruments for showing the hour, and those they have are so poor that you cannot trust them". The same emperor who set up a clock-making operation referred to the clocks his children had as something "to play with". And, unlike the Japanese, the Chinese did not develop a clock modified to fit their own traditions. Furthermore, dependence on European imports continued.

Was all this simply a result of the cumbersomeness of the Chinese political system combined with a fierce desire not to admit that a non-Chinese society had something better? Missionary belief that European technology proved a better religion did not necessarily set up a smooth interaction. Unquestionably some Chinese scholars, by the 17th century, were arguing, not always consistently, that: water clocks were at least as good as mechanical clocks; that it was actually the Chinese who first invented mechanical clocks, so the West were merely copying; and that Western devices did not keep good time, because the spring mechanisms constantly loosened. "But these clocks, which are so difficult to make and break down so easily, aren't they really nothing better than pure waste?"

And, finally, what had been at least a somewhat complex relationship, between European clock advocates and Chinese officialdom, broke down in the early 18th century. The Catholic Church established a new rule that missionaries must dress and behave like Europeans, instead of adapting to local ways as Ricci had done successfully; in response, the Chinese radically curtailed missionary activity. Interaction languished, aside from ongoing trade. So when a British delegation visited the emperor at the end of the century, hoping for better trade and bringing clocks, among other things, as tokens of good will, the emperor rejected the overtures noting among other things, "We have never set much store on strange and ingenious objects, nor do we need any more of your country's manufactures."

Even here, however, it pays to be cautious: though the Chinese government and a sense of cultural superiority were powerful factors, they did not monopolize Chinese interests. Shiploads of European clocks were being exported to China by the late 18th century and, while haggling over price could complicate a deal, clearly a number of Chinese, mainly upper class but possibly including some members of the active business community, were buying clocks. What had not happened, admittedly, was a national incorporation of clockmaking.

Official Chinese hesitation over clocks increased Europeans' disdain for the giant empire, as they increasingly judged societies by their openness to technological innovation. It left many Chinese practices, including imposing an interminable wait for anyone seeking an (untimed) audience with the emperor, notoriously inexact and wasteful. And it leaves historians still wondering whether the Chinese must be condemned (retrospectively) for failing to adapt more quickly, or whether there is a more benign interpretation.

Reactions to European mechanical clocks varied. Factors included: range of preexisting contacts with Europe; religious connections; adequacy of existing instruments for determining time; artisanal culture; and, possibly, at least in China's case, extent of civilizational defensiveness. There was wide interest, particularly among the upper classes, but also a great deal of fuzziness between fascination with artistic and mechanical gimmickry and actual time measurement.

It is important not to read too much into these regional variations. Clocks remained inaccurate (as well as expensive): it was not necessarily reactionary not to develop an overwhelming interest. Europeans themselves often valued clocks mainly for their novelty value, just as some of the Chinese or Ottoman elites did. And, by 1750, it was not clear that any civilization was losing ground simply because it did not replicate European levels of clock use. The Chinese manufacturing apparatus, for example, continued to purr along nicely. Only in the 19th century, as clock time interlaced with industrialization, would the results of differentiation become more obvious.

There is, however, one final point to cover, before turning to industrial uses of time directly, a point that returns to Western Europe as well as, increasingly, the British colonies in North America. Further technological change and, even more, a wider public embrace forged more intense connections among clock time, economic life and, to some extent, life in general by the 18th century.

Fascination with Time: The 18th Century

Two or three developments mark the 18th century, in Western society, as something of a culmination point of the various trends that had been developing since the advent of the mechanical clock. The uses of clock time measurably expanded, with some new technology attached; but, perhaps more important, a larger cultural fascination with time began to emerge, at least in some key social groups.

It is also interesting to note that it was in the 18th century that the idea of the century itself became a more definite part of Western culture (with some precedents from earlier Christian chronologies). A number of people exchanged messages and greetings to herald the advent of the 18th century itself, and a judge in Massachusetts even wrote "Verses upon a New Century", which he had the town crier read in public. Thinking in terms of centuries might be seen as part of a growing mentality that thought in terms of time as divided into regular, systematic categories.

Navigation

As European ships began regularly to cross the oceans, from the 16th century onward, they encountered a huge navigational challenge: it was impossible to determine longitude with any accuracy, with available equipment. Latitude was not a major problem: sighting the position of the sun at noon, or the position of the Polaris star, allowed reasonably good calculation of a ship's location relative to the equator. But longitude was complete guesswork. The resulting ignorance caused frequent shipwrecks, particularly in the Pacific but elsewhere as well. The crisis was widely recognized.

Various European governments, including Spain, tried to encourage a solution to the problem. By the 17th century the issue engaged leading intellectuals, including Isaac Newton and Christian Huygens (of clock fame). It was clearly understood by this point that calculating longitude required knowledge of the exact time, but still there was a gap to be bridged. Early in the 18th century the British government (aware that its ships were increasingly plying the Pacific as well as other oceans), offered a major prize for innovation. And, partially in response, several imaginative artisans came up with crucial inventions.

One device was the sextant, anticipated by Newton but actually developed by John Hadley, a British mathematician, and Thomas Godfrey, an American optician and inventor, around 1730. This was a doubly mirrored instrument that allowed simultaneous sighting of two astronomical objects, thus determining their visual distance from each other. This was a huge step forward, but for accuracy it depended on knowing the exact time, which could be compared to an astronomical chart. If star X and star Y are in this relationship at 7 PM, we know where we are on a north–south axis – we know our longitude – but only if we in fact know that it is 7 PM.

And here a slightly earlier invention filled out the puzzle. John Harrison, a self-taught British carpenter, developed a vacuum chamber to contain a clock, which in turn would keep accurate time even amid changes in temperature and during the heaving motion of a ship at sea – and it was Harrison who collected the British government prize, in 1714. The device was steadily improved during the 18th century and given the name "chronometer" by John Arnold, a clockmaker, who devised a sensationally accurate version as a watch in 1779. These devices were capable of keeping accurate time with a margin of only three seconds per day – more accurate than existing clocks even on land.

This achievement constituted possibly the most important step forward in clock-making ever recorded. Harrison himself, commenting on a version he had improved by mid-century, noted:

> I think I may make bold to say that there is neither any other Mechanical or Mathematical thing in the World that is more beautiful or curious in texture than this my watch or Timekeeper for the Longitude….and I heartily thank Almighty God that I have lived so long, as in some measure to complete it.

Chronometers and sextants quickly became standard devices on ships of all sorts, not only in Britain, supporting a burst of exploration in Pacific Oceania and the American northwest. They played a vital role in extending the domains in which accurate time was both vital and available, and in improving the accuracy of timepieces even in other settings.

Problems persisted in more general usage. Writers in the 18th century still complained about how each person's watch seemed to yield a different time – making it easy to distrust one's own instrument. A mid-century dictionary warned readers against excessive confidence: "One should put little credence in the common notion most people have of the accuracy of their watches." But conditions were improving, and the quote itself suggests how a growing number of Europeans were beginning to seek precise time, personally delivered, in their daily lives.

Range of Application

More accurate time and greater reliance were reflected of course in the growing array of activities subject to clock time. Navigation itself was a key example. It was also in the mid-18th century that British stage coaches began to project their competitive, time-conscious schedules, with a coachman provided with a watch to make sure he kept up the pace.

Also in the 18th century, as manufacturing continued to expand, a few industrialists began to experiment with clock time as a means of imposing greater efficiency on workers. Most production continued to occur in homes or small shops, with individuals or family groups determining schedules on their own. But a few businessmen, particularly in Britain, began to cluster production in factories, where workers could be supervised by others. Here, workers could be put on the clock, with factory bells indicating when everybody was to show up for the day, when a lunch break started and ended, and so on. Early efforts might encounter rough going. One factory owner found that he had to ring the end-of-lunch bell several times, because his workers were still accustomed to make their own decisions about when they were ready to work. But the system began to catch on, and it established precedents that would become far more important in the following century.

Availability

For many people, even at the end of the 18th century, price was a more important problem in timekeeping than accuracy was. But here too, there was movement, though not yet a revolutionary breakthrough. A growing number of artisans and self-taught tinkerers began to experiment with cheaper models. We have seen the efforts, in the American colonies, even to use wooden parts, to save on expense.

More important still was the steady increase in the number of shops that produced clocks and watches. Britain long led the way. By the late 18th century, hundreds of master craftsmen, employing thousands of outworkers, were manufacturing tens of thousands of watches a year. Between 1775 and 1800, the output totaled between 150,000 and 200,000. Much of this was for export of course – watches were still mainly for an elite, both domestic and foreign, save for categories like navigation where they had become standard fare. Still, there was no question about a steady increase in supply. Manufacturing was increasingly broken down into specialized units, with domestic workers churning out springs or dials or cases for later assembly in a finishing shop. And prices did go down steadily, from the late 17th century onward.

British production was increasingly rivaled by operations in Switzerland, where similar production principles were applied. Many Swiss watches were even smuggled into Britain, because of their quality and price. Competition among European centers – France was another leader (though Germany had fallen behind because of religious turmoil in the 17th century) – also had its effect on price and availability. And in the final decades of the century, entrepreneurs in the Swiss countryside began introducing further innovations in the production process, using machines to stamp out more of the parts and assembling up to a hundred workers in factories that could turn out over 40,000 units a year – some of them relatively crude but unquestionably cheap.

This was still, at the end of the 18th century, a skilled craft operation overall, in terms of the ultimate product, with a maker's individual brand proudly stamped on each final unit. But volume and affordability were evolving, and this both reflected and encouraged the growing uses of clock time.

A Culture of Time

It is impossible to say how many people, in Western Europe and the British colonies of America, were routinely time-conscious by the later 18th century. The growing eagerness to buy clocks and watches, many of them shorn of luxury trappings, and the expanding range of activities in which time was important is clearly noteworthy. The emergence of figures like Benjamin Franklin, eager to advise a wide readership of the importance of paying attention to time and using it efficiently in daily life, surely both reflected and encouraged wider interest.

And there is another set of indicators as well, at least strongly suggestive. Google Ngrams allow a rough estimation of the relative frequency with which a topic was being discussed in an array of published materials at any given point since 1500. It is important to emphasize the crudeness of this evidence: not all publications have been digitized, and the fact that a topic was changing in frequency does not prove that many people were actually altering their values. And of course the Ngrams' utility depends considerably on literacy, though this was becoming more common in the Western world by the 18th century.

Still, with all due caution, the Ngram data are suggestive. Relative interest in "time" as a subject accelerated rapidly in the 18th century in the English-speaking world, reaching its highest point before or since (see Figure 4.1). In French, references to "l'heure" reached a similar mid-century peak, though in this case it would crest again in the later 19th century. More and more people, or at least more and more readers, were being exposed to various discussions of time. (Ngrams for discussions of clocks followed a related but slightly different pattern, peaking a bit earlier, in the later 17th century, though maintaining a fairly high level into the 18th.)

This is not all. A similar pattern prevailed in English for the word "punctuality", where again a high point surged in the 18th century. Concern about idleness, an old staple in many cultures, may have been shifting along with the wider interest

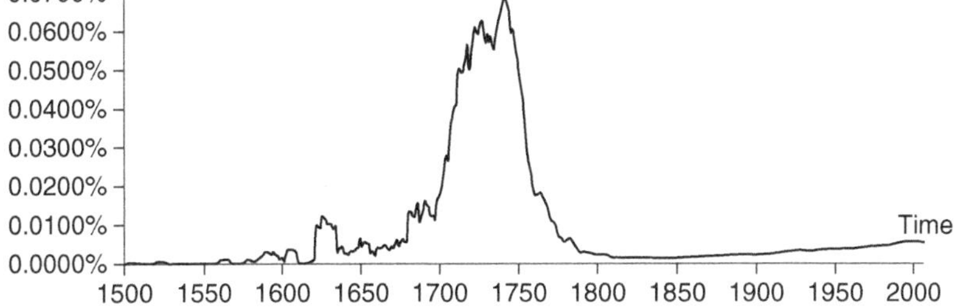

Figure 4.1 Frequency of the word "Time" in English, 1500–2008, Google Ngram viewer, accessed August 15, 2019.

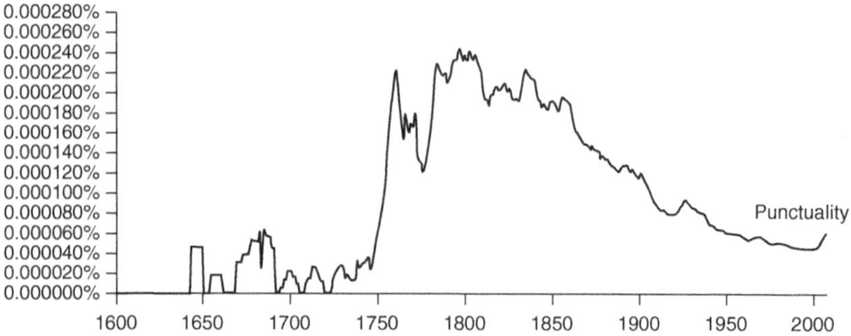

Figure 4.2 Frequency of the word "punctuality" in English, 1600–2008, Google Ngram viewer, accessed August 9, 2019.

in clock time, making – as Ben Franklin was urging – timeliness a new virtue (see Figure 4.2).

Further, it was in the 18th century that the phrases "regular as clockwork" or "just like clockwork" began to gain currency in the English language. The word "clockwork" had been introduced early in the 17th century, reflecting the growing awareness of the operations of mechanical clocks. But the notion that clocks could be referred to more generally as a model of rational, predictable operation was an innovation, again suggestive of a new interaction between sense of time and the wider culture.

Many intellectuals were even extending the clock image to the religious realm, arguing that clockmaking was the best way to describe God and his creation of a smoothly functioning, relatively rational universe. Even musicians picked up the theme. It was in 1796 that the composer Haydn issued his "clock symphony", with the second movement featuring a clock-like beat. Not just the uses of time, but the consciousness of time were expanding.

A Rural Case

One final data point, though it intrudes into the early 19th century, helps suggest the evolution. Clocks and watches began to reach the countryside, though – as had been true earlier for the urban upper class – they generated a concern for time only after a generation or so.

In the Hudson River valley, in New York state, a rural area but exposed to increasing trade, clock sales began to accelerate from the 1780s onward. Traveling salesmen eagerly pressed local farmers to take the plunge, noting that the locals valued aesthetics but were also concerned about price. Early references to clocks and watches emphasized their contributions to social status: thus a village paper praised clocks and watches for their role as "embellishments", constituting "a very valuable appendage for the gentleman." For in the 1780s, knowledge of hourly time was not very important in this largely self-sufficient region: most transactions were local and personal and, while seasonal change was carefully noted, greater detail was not necessary. Local almanacs printed sheets that could be carried in a pocket, indicating the length of days, by season, throughout the year. Little more was necessary – except for contributions to consumer status.

By the early 19th century, however, as outside merchants came in to buy and sell, knowledge of clock time became more significant – as a basis for arranging appointments to discuss transactions; and other local gatherings, like meetings of temperance societies, began to be scheduled by the clock. Increasingly, even small villages began to erect public clocks and, while arguments for these clocks sometimes referred to rivalries with other communities, the main point was "to have a town time that could be relied upon". The change – the emergence of a new and widely shared time sense, had two bases: first, the shift in the economic environment, requiring new kinds of scheduling and, increasingly, a new pace of work; but, second, the experience, over several decades, of becoming accustomed to a timepiece that initially may have been purchased simply for show. The result, unquestionably, was a new internalization of time, spreading from local businesses to the farming population; but it developed somewhat circuitously.

This history of clocks and time between 1300 and 1800 suggests a reasonably steady evolution that, ultimately, added up to an arguably dramatic change. To be sure, the fascination with mechanical clocks and their implications for daily life might spur a more sudden shift for particular individuals. It was already in the 15th century that the Italian artist Alberti – the same one who urged fathers to record the times of their children's births – talked about how he organized each day in terms of time. "What am I going to do today? So many things: I count them, think about them, and to each I assign its time"; it is vital to "watch the time, and assign things by time, to devote oneself to business and never lose an hour of time. ….He who knows how to make use of time, he will be lord of whatever he wants." Here, clearly, was a "modern" time sense, in terms of the level of involvement independent of religious ritual.

But for most urbanites, the evolution of clock use and time awareness was more nuanced, undercutting some of the most ambitious claims about the gap between a modern West and its more traditional past. Many people continued to entertain different patterns of time depending on the issue involved. The accuracy and availability of timepieces shifted gradually, and a conversion from novelty item or consumer display piece, to routine use, was gradual as well. All of this makes sense, in terms of the complexity of clock technology and a common human impulse to avoid too much sudden disruption in personal rhythms. But it offers a more complex picture than some revolutionary imagery suggests.

The same complexity, we have seen, applies to the comparisons between Western Europe and other regions. Important differences did emerge, but they were sometimes subtle. Non-European elites did not systematically reject innovation, but on average the nature of their embrace was different.

Finally, developments by the late 18th century were only part of the modern story. They unquestionably helped prepare the way for further change. The shifts in stage coach scheduling, that allowed for new use of clock time, made adjustments to the time demands of the railroad less jarring – though the impact of rails on time were not as straightforward as might be imagined. Experiments with applying clock time to daily work – from the use of the bells of the clock towers by municipal authorities to the first experiments with factory bells – clearly suggested future opportunities. Whether the real revolution had occurred by 1800, with the aftermath mere extension – or whether, for most people, the revolution was yet to come, is not easily determined. The histories are connected but, in 1800, a lot of fundamental, sometimes quite painful, change was still in the future.

Notes

1 The word "watch" itself was introduced into the English language only at the end of the 16th century, though there may have been one reference earlier. "Watch" as a term already existed, to emphasize waking up or staying up at night to keep guard or participate in revelry. We have already seen the importance of timing devices for regulating the hours of night guard watches. The word watch itself now spilled over into the small timepieces by the 1580s, from earlier emphasis on the importance of small clocks "to wake up sleepers".

2 Then in the 1650s the pendulum was introduced for the larger clocks, providing much more regular gear motion than mere weights could offer. The great Italian scientist Galileo, early in the 17th century, had studied the motion of swinging objects and may have projected a pendulum idea. But the development was brought to fruition by a Dutch mathematician and scientist, Christian Huygens, who completed the first pendulum clock in 1657. Pendulums could be used either with weights or springs, and again the benefits quickly showed in greater accuracy. Around the same time the escapement mechanism was improved as well.

Further Reading

Two major studies provide superb introductions to this period, though they also present some nuanced disagreements:

Landes, David S. *Revolution in Time: Clocks and the Making of the Modern World*. (New York: Barnes & Noble Books, 1998) is a vital study based on wide empirical research.

Dohrn van Rossum, Gerhard. *History of the Hour: Clocks and Modern Temporal Orders*. Translated by Thomas Dunlap. 2nd ed. (Chicago, IL: University of Chicago Press, 1998) covers a still wider set of research findings.

On big claims about the modern time sense:

Bilfinger, Gustav. *Die Mittelalterlichen Horen Und Die Modernen Stunden: Ein Beitrag Zur Kulturgeschichte*. (Charleston, SC: Nabu Press, 2013).

Cipolla, Carlo M. *Clocks and Culture: 1300–1700*. (New York: W. W. Norton & Company, 2003).

Crombie, A. C. *The History of Science from Augustine to Galileo*. (New York: Dover Publications, 1996).

Mumford, Lewis, and Langdon Winner. *Technics and Civilization*. (Chicago, IL; London: University of Chicago Press, 2010).

Quinones, Ricardo J. *The Renaissance Discovery of Time*. (Cambridge, MA: Harvard University Press, 1972).

On longitude:

Howse, Derek. *Greenwich Time and Longitude*. Revised edition. (London; Wappinger's Falls, NY: Philip Wilson Publishers, 2003).

Gould, Rupert T., and Jonathan Betts. *The Marine Chronometer: Its History and Development*. (Woodbridge: Acc Art Books, 2013).

Quill, H. *John Harrison: The Man Who Found Longitude*. (New York: Humanities Press, 1966).

Sobel, Dava. *Longitude: The True Story of a Lone Genius Who Solved the Greatest Scientific Problem of his Time*. (London: Bloomsbury, 1995).

On clock production:

Jaquet, Eugène, and Alfred Chapuis. *Technique and History of the Swiss Watch*. Revised edition. (London, New York: Spring Books, 1970).

Weiss, Leonard. *Watchmaking in England, 1760–1820*. (London: Robert Hale Ltd, 1982).

On changes in the arrangement of family meals:

Shammas, Carole. *The Pre-Industrial Consumer in England and America*, 2nd ed. (Los Angeles, CA: Figueroa Press, 2008).

On the advance of the idea of centuries:

Stearns, Peter N. *Millenium III, Century XXI: A Retrospective on the Future*. (New York: Perseus, 1998).

On the West–rest relationship:

Adas, Michael. *Machines as the Measure of Men: Science, Technology, and Ideologies of Western Dominance*. (Ithaca, NY: Cornell University Press, 1990).

On the Middle East:

Kurz, Otto. *European Clocks and Watches in the Near East*. (London: Warburg Institute, University of London, 1975).

Uluengin, Mehmet Bengü. "Secularizing Anatolia Tick by Tick: Clock Towers in the Ottoman Empire and the Turkish Republic." *International Journal of Middle East Studies* 42, no. 1 (2010): 17–36.

On Russia:

Lucas, Gavin. *The Archaeology of Time*. (London; New York: Routledge, 2005).

Miller, David B. "Monumental Building as an Indicator of Economic Trends in Northern Rus' in the Late Kievan and Mongol Periods, 1138–1462." *The American Historical Review* 94, no. 2 (April 1, 1989): 360–90.

In developments in Japan:

Frumer, Yulia. *Making Time: Astronomical Time Measurement in Tokugawa Japan.* (Chicago, IL: The University of Chicago Press, 2018).

Mody, N. H. N. *Japanese Clocks.* (North Clarendon, VT: C.E. Tuttle Co, 1967).

See also Needham, Joseph. *The Hall of Heavenly Records: Korean Astronomical Instruments and Clocks, 1380–1780.* (Cambridge; New York: Cambridge University Press, 2004).

On colonial America:

Bruegel, Martin. " 'Time That Can Be Relied Upon.' The Evolution of Time Consciousness in the Mid-Hudson Valley, 1790–1860." *Journal of Social History* 28, no. 3 (March 1, 1995): 547–64.

Gibbs, James W. "Thomas Jefferson's 'Wonderful Clock'." *Bulletin of the National Association of Watch and Clock Collectors.*, n.d.

Hall, Dennis R. "The Watch as a Cultural Icon." *International Journal of Culture Tourism and Hospitality Research* 2, no. 1 (March 2008): 5–11.

Hering, D. W. "Geographic Distribution of Early Clockmaking in America." *American Collector* (1937).

McCrossen, Alexis. *Marking Modern Times: A History of Clocks, Watches, and Other Timekeepers in American Life.* (Chicago, IL: University of Chicago Press, 2016).

Smith, Mark Michael. *Mastered by the Clock. Time, Slavery and Freedom in the American South.* (Chapel Hill: University of North Carolina Press, 1997).

On Latin America:

O'Hara, Matthew D. "The History of Time in Colonial Latin America." *History Compass*, 11 (2013): 77–88; the article contains further references.

On China:

Baichun, Zhang, and Tian Miao. "Joseph Needham's Research on Chinese Machines in the Cross-Cultural History of Science and Technology." *Technology and Culture* 60, no. 2 (June 14, 2019): 616–24.

Landes, David S. *The Wealth and Poverty of Nations: Why Some Are So Rich and Some So Poor.* (New York: W. W. Norton & Company, 1999).

5 Imposing the Clock

Uses of Time in Industrial Society, 1800–1920

From the factory rules of the Benck textile company in eastern France (1843):

> Every worker who accepts employment in any work site is obligated to read these rules and to submit to them. No one should be unfamiliar with them. If the rules are violated in any work site, the offenders must pay a fine according to the disorder or damage they have caused.
>
> The work day will consist of twelve hours, without counting rest periods. Children under twelve are excepted. They have to work only eight hours a day.
>
> The bell denotes the hours of entry and departure in the factory when it first rings. At the second ring every worker should be at his work. At quitting time the bell will also be sounded when each worker should clean his workplace and his machine (if he has one). It is forbidden under penalty of fines to abandon the workplace before the bell indicates that the work site is closed.

This chapter and the next cover what is sometimes called the "long" 19th century – from the late 18th into the early 20th. Like the previous period, these decades saw a number of crucial developments in the history of time – though they have, on the whole, won less attention from specialists in the subject. The big foci are, first, how industrializing societies both needed and facilitated new uses of time – the topic of this chapter, centered on Western Europe and the United States; and, second, how increasing global trade and contacts, largely under Western sponsorship in this imperialist age, generated change and resistance regarding the nature of time in other regions – the topic of Chapter 6.

As in the early modern period, developments centered particularly on clock time during the day and, increasingly, the night, though there were some global changes in accepted calendars. But even in the West, where the Gregorian calendar remained the common currency, some important shifts occurred in the conception of the week – most particularly the emergence of the modern weekend.

As to clock time, attention shifts away from basic technological advances, though there were a few, to the abundance of possibly painful adjustments in the ways ordinary people encountered time – at work, at play, in childhood, even in routine manners. While real questions remain about the extent of conversion to a modern time sense by 1800, for the bulk of the population, there are no doubts about 1900: clock time was everywhere. To be sure, most people, even in the West, did not yet have personal timepieces. And there were debates around the edges about the importance of great precision. It was around 1900 that a German military leader opined that "practical life rarely demands a punctuality reckoned in minutes", noting that schools and courts often actually set clocks back so that attendees would not be declared late. But his nonchalance was answered by an official from the Berlin observatory, who insisted that timelines to the minute were a matter of basic courtesy. And even the general would not have disputed the fact that minutes, and certainly hours, half hours and even quarter hours had become steadily more important during his lifetime.

The key question is not the spread of the importance of clock time and devices to measure it, but the mixture of voluntary and compulsory incentives for change. It is abundantly clear, historically and as a matter of common sense, that clock time is not natural, but of human devising. As it became more ubiquitous in the Western world, could it be assimilated fairly readily or was it an unwelcome imposition? Answers depended greatly on social class – and related experience with clock time before the industrial revolution – and sometimes even on gender.

Technology and Production

Not surprisingly, given existing trends before 1800, clock and watch technology steadily improved, generating greater accuracy. By the later 19th century, national governments began to provide testing and measurement programs by which timepieces could be evaluated. Switzerland set up this kind of operation in 1876, France in 1885. A testing center in London, initially located in Kew Gardens, examined 15,301 different brands of clocks and watches in the three decades after 1884. In the United States, often a bit delayed in the arena of government initiatives, the Bureau of Standards began testing watch types in 1914.

Further, amid growing national and international competition for the production of timepieces, inventors and manufacturers were eager to introduce better or newer products. In the first half of the 19th century, for instance, American firms often claimed innovation, around new types of clock design. One example: the very popular banjo clock, so called because of its distinctive shape, won considerable appeal. Production of what were called "shelf clocks", two or three feet high, was another important American development. Also in the United States, the manufacture of wooden clocks, previously popular as a money-saving alternative, phased out by the 1830s. A new entrepreneur, initially trained as a carpenter, introduced an affordable all-brass model that simply beat them out. Chronometers were also steadily improved, now for use on trains as well as ships.

Attention certainly focused on clocks and watches for personal, home, or office use, and not on the great public clocks that had been so important in the previous period. Tower clocks were still produced. A major venture in New York, in 1880, ten years in the making, featured a device 18 feet high, which indicated the time in various places around the world, the date, the day of the week, and the motions of sun, moon, and the major stars. On the hour, a figure of George Washington rose, extended his right hand to present the Declaration of Independence (even though he never did this in fact), and then a series of former presidents paraded out to salute the country's Father. And, even aside from special projects of this sort, huge clocks remained important, most obviously in the new genre of railway stations. At the same time, intense urban growth was generating cities that were much too large to be within sight or sound of a major public clock.

As a result, increasingly, this was a period when smaller devices took center stage. And there were some interesting technological developments in the category, beyond artistic design. Stem-winding pocket watches increasingly replaced the traditional product wound with a key – a major practical advance, since it had been so easy to lose the key. The formal invention of the stop watch, in France in the 1820s initially to check the speed of the king's race horses, began to win wider popularity in human sports by the end of the century. And while earlier clocks had often been adapted to generate a wake-up bell (even some water clocks had offered this function), it was in 1847 that a French inventor first patented an adjustable and transportable alarm clock. The alarm clock became popular enough that Fanny Brice, the American comedienne, could tell the following joke, listed as one of her five best, on stage in 1891: A man wakes up mid-morning, clearly going to be late for work. "What happened to the alarm clock?" he asks his wife anxiously. "It went off at 6", was the answer. Apparently funny at the time.

With all this, the big news in the long 19th century was not in clock technology, but in the production systems deployed – and therefore in the sheer quantity of items available for sale. Headed initially by American manufacturers between about 1820 and 1850, most of them former artisans, producers increasingly applied the great principle of interchangeable parts, first introduced in Britain and the United States mainly to accelerate the production of guns, to the making of clocks and watches.

A British observer in 1880s, while insisting (correctly) the American factory products were inferior to the crafted but pricey British models, described the process this way: In United States centers, "watches were made by machinery, so that every piece will fit every watch of the same pattern". By making manufacture as automated as possible, American entrepreneurs could pay their workers higher wages but still offer lower prices for their products. Earlier, one of the American factory pioneers, Edward Howard, described how he turned to use of machinery over skilled craft labor. Himself from a craft background, he observed how slow and painstaking the traditional methods were, often generating only a few hundred clocks or watches a year. He decided to take the plunge for a newer methodology: "The development of this plan to use automatic machinery was the result of long thinking."

And it paid off. A number of American factories, disproportionately located in Connecticut, began to spur national output. Britain was slow to respond, and lost ground in the industry despite high quality products; despite some local objections, American imports began to cut into the market even at home. But France adjusted to some extent (by 1900 accounting for about 12% of world output), and Swiss manufacturers, initially challenged, turned to more automatic machinery with a vengeance, claiming greater numbers than the United States could boast by 1900. The Swiss position was bolstered by a tradition of high quality and by wages that were somewhat lower than American levels. And manufacturing technologies themselves steadily improved: by 1880 a single American factory worker could account for up to 10,000 watch gears in a single day.

The result was a real explosion of timepieces. In 1800, worldwide, about 400,000 watches were produced per year. In 1875, the figure had risen tenfold, to 2.5 million. And this was just a start. By 1890 Switzerland alone was exporting 5 million watches each year (including to the United States), and by 1910 the figure had risen to 10 million, growing about 5.6% per year, a huge surge. The United States was producing several million watches a year in the same period, focusing primarily on cheaper models, while serving as the tenth-largest national market for Swiss imports.

These big numbers must be handled carefully. Swiss watches were sold all over the world, and the world was a big place. Considerable export activity centered on eastern Europe and parts of Asia. Still, production and sales figures suggest that a large minority of at least the urban population of Western Europe and the United States (and the urban population was nearing half of the total in this region), owned a watch or a clock or both. An annual output of, say, 20 million units a year by 1910, of a product that, while susceptible to breakdowns, usually lasted for a while, suggests considerable popular access. And prices, while not yet at working-class levels, had fallen well out of the luxury category as well. A good but cheap watch in the United States cost about $12, or what in current dollars would be about $140 – not out of sight for many people.

The main point is clear. What had been a fairly rare upper-class item during most of the early modern period had spread to the bulk of the middle class (at least in terms of one watch or clock per family) and somewhat beyond. This change reflected the new need to know what time it was – this was no longer primarily a conspicuous consumption category, though luxury was still available for the few. The change also encouraged greater interest in knowing the time. Both facets – need and interest – can be explored more fully.

The Cultural Context

Along with the growing availability of timepieces, several cultural outlets began to highlight the importance of time-based punctuality in new ways. This factor built on developments in the early modern centuries – like the advice emanating from Benjamin Franklin. And it centered on clearly middle-class values, raising questions about its impact on other groups. But its insistence was undeniable.

A new genre of self-help books emerged, with figures like Horatio Alger in the United States and Samuel Smiles in Britain, bent on claiming that good character and hard work, not social rank or even genius, were the basis for success in life. Samuel Smiles' famous book, titled simply *Self-Help,* filled with examples of careful use of time. Thus the novelist Walter Scott, who started work promptly at six in the morning: "Punctuality was one of the most carefully cultivated of his habits." Diligence, perseverance, industriousness all centered on awareness of time: "not a moment to be lost"; "meaningful employment of fragments of time". All of this, by the mid-19th century, was fairly predictable, given the precedents available from the 18th century, but it reached an audience eager to learn how to do better in life. Smiles' book sold 250,000 copies in the second half of the century, appealing not only to middle class readers but also to many artisans who were trying to figure out how to better themselves as traditional craft skills were increasingly challenged.

More interesting than the self-help genre was the importation of advice about time into books on good manners. Here was a type of literature that had developed in Europe since the Renaissance, but that had not previously paid much attention to issues of time. And, in fact, many social gatherings, like the famous French *salons* of the later 18th century, had been fairly open-ended, with people coming and going over an extended period.

Not so for the new world of the 19th century. Dinner parties, for example, should be carefully timed, according to the new, widely popular etiquette advice. The hostess should be available at least 15 minutes before the appointed hour. And guests, too, should show up promptly. "It is of the utmost importance that all of the company be punctual, arriving 10 to 15 minutes before the appointed time." Tardiness of even ten minutes was "a serious offense, which no one should be guilty of". On the other hand, arriving too early was also impolite: good manners involved keeping careful track of time. And while meals themselves were not rigorously timed, hosts and guests were urged to make sure that eating and conversation extended at least to 30 to 60 minutes.

Similar recommendations, in the etiquette books, applied to attendance at business meetings or medical appointments. Here too, arriving ten to 15 minutes early assured that a person was eager "to appear professional, unruffled and prepared", in addition to showing respect for others. Weddings were another occasion where punctuality was called for. And several manuals also emphasized the importance of "being early" when traveling.

The new juncture between manners and timing was an important development in extending cultural pressure on daily behavior. Again, the genre was primarily middle-class, though it had explicit implications for the responsibilities of servants. In a period intensely concerned with living up to standards of "respectable" behavior, the emphasis on punctuality could make a deep impression.

Rounding out the cultural side was the increasing treatment of time and clocks in popular literature. Charles Dickens, for example, writing in the mid-19th century, was fascinated with clocks, and frequently featured them (both house clocks and the bells of public clocks) in his stories. We will see that references were increasing built into materials aimed at children as well.

Time and Work

It was the workplace where the concern for living by the clock had the most obvious impact on ordinary life. We have noted that efforts to regulate work by the clock go well back in time, developing particularly during the early modern centuries in Western Europe. Many local governments sought to define beginning and ending points by the time of day, and sometimes lunch breaks as well: urban clocks and bells were good enough to allow this. Disputes with employers, who tried to argue that a given worker had not put in all the necessary hours and so should be docked in hourly pay, or efforts to win additional pay for hours beyond the norm, suggest how hours and work could be linked. Finally, the example of some British factories that tried to shift to hourly time in the later 18th century, with bells calling workers to assemble, correctly suggests that the time–work relationship was beginning to develop still further.

Still, in 1800, and even in urban Britain, most work was not organized around the hour. Most manufacturing still occurred in small shops, where the people involved set their own pace. A great deal of manufacturing production still paid by the piece, rather than time spent. That is, a worker generated an agreed-upon length of cloth, or a set number of scissors or shoes, and was paid accordingly – no matter how long it had taken. The timing was up to the individual worker. Piece payments helped merchant-manufacturers control their costs, in a situation where they had little detailed control over the daily operations of the labor force.

But with the rise of factory industry and the application of new, often power-driven machinery, all bets were off. There were three interlocking reasons increasingly to shift to a system of work and pay based on the hour, and with greater detail than ever before – reasons at least from the manufacturer side, with the results pressed on the growing masses of factory workers.

#1: With increasingly elaborate machinery, often organized around a steam engine, groups of workers needed to be assembled to start work at a common time. Machines required teams of workers, not single operatives. And many workers, in growing cities, lived at some distance from the plant. Coordinated timing requirements seemed essential.

#2: The whole reason behind factory production centered on increasing output, often very rapidly. If workers were paid by the piece, there might be endless quarrels over what the rewards for greater output should be, and to whom – worker or manufacturer – they should go.

#3: The main reason that output expanded so rapidly rested on the power of the new machinery, not (at least according to employers, who were very conscious of the risky investments they had made to get the machinery in the first place) workers themselves. A pay system seemed essential that would not give workers too much benefit from the productivity of equipment for which they were not responsible – in some cases, in the rigors of early industry, that might give them no benefit at all. Hourly pay, obviously, might be held steady or even decreased as volume went up.

And all of this took shape in an environment where, as we have seen, many business people were already suspicious of popular working habits, with what they claimed were wasteful impulses to "loiter" and fritter away time. How much this kind of prejudice affected new factory approaches to time, and how much resulted more objectively from the demands of new machinery, is not easy to sort out.

But there was no question about the result. Increasingly in the early factories, organization of work by the hour became a standard feature of industrial life. In turn, the imposition of strict time requirements became one of the most distinctive qualities of modern work – beginning in the factories, but soon spreading more widely.

Time Policies

Most factories, accordingly, made time a centerpiece of the now pervasive "shop rules", like those of the Benck company cited at the start of this chapter. Formal efforts to regulate punctuality dot the 19th century, from Britain to Russia. Employers clearly believed, and possibly with some justice, that workers' sense of time had to be carefully regulated and, in many cases, corrected. Some shop rules, in addition to specifying the starting hour (usually with a bell to warn, and then another one 15 minutes later to seal the deal), made it clear that the factory door would be locked once the day had begun. Workers who were late would lose at least half a day's pay, and might be fined in addition before they could be readmitted. Porters were employed at the door by way of enforcement, and supervising foremen – a new lower management slot – kept careful track of workers' presence as well.

By the 1830s, in many factories (and particularly in factory towns), time controls were supplemented by a new device, better than a bell. Steam whistles, initially developed for locomotives, allowed a piercing sound to be carried more widely, alerting workers that their day was about to begin.

The importance of clock time was also highlighted by the growing movement to base salaries on hours rather than piece or product, though there were ongoing discussions about this issue by employers and workers alike. Increasingly, the working class was partly defined by their subjection to hourly rates of pay, which further highlighted the importance of timing units in their lives.

Additionally, clock time was increasingly extended into the night. Obviously 12- or 13-hour days assumed night work and, during winters, smoky whale oil lamps maintained this system with no regard for season. Later, factories would introduce gas lighting, common in Britain by the 1840s, and then after 1880 electricity. In some cases, multiple shifts of workers were organized through an hourly schedule.

And in all this the imposition of the clock was frequently accompanied by bitter blasts against the laziness and immorality of the working class: enforcement of strict hours was not just a matter of business efficiency but a blow toward uplifting a dangerously corrupted group of people. Here is a characteristic lament, from Britain in 1823, concerning what workers did when they were allowed "to do as they please" for some part of the day. "We shall often see them just annihilating those portions of time. They will for an hour, or for hours together…sit on a bench, or lie down on a bank or hillock…yielded up to utter vacancy and torpor." Time discipline, in the minds of the more smug employers, was doing everyone a favor.

There is no question about what happened: work became more closely time-regulated, in ways that coincided with the demands of new machinery and also fit into the middle-class moral sense, filled with rhetoric about the importance of using every moment wisely. Of course, as always, there was variety: what happened in the factories could not be imposed, for example, on construction sites, where work might begin and end by the clock but with a much less regulated pace during the day itself. And even some factories – for example, in the British pottery industry – had not managed to impose rigorous time as late as the 1920s.

Nevertheless, the trend was clear, and the rigors of the clock were very real for growing numbers of workers. And this yields a second question: about how workers themselves responded to the new regime of clock time – granting always that individual personalities might vary where punctuality was concerned.

There are several variables. First, at least in the early decades of industrialization, a certain number of employers cheated. They set up time schedules on paper, but then sought ways to force workers to stay on the job longer than stipulated (longer than the 12–13-hour days already officially required). Many employers, in several countries, were accused of moving the hands of the factory clock, to make it seem earlier than it was (thus compelling workers to stay longer than required), or shortening lunch hours. In one striking case in Britain, the employer and his son were the only ones on the work site who had watches: most workers had no idea what time it was. And when one worker showed up with a watch of his own, it was confiscated. All this is interesting not just for the shocking chicanery, but also in indicating that many workers were in fact familiar with clock time – if they had a fair shot at a clock.

A second variable casts back to the interpretive dilemma discussed in the previous chapter. Some workers may have been quite familiar with clocks before the factory system widely took hold. They might resent the number of hours required, or the intrusive supervision, but the idea of paying attention to time at work would not have seemed shocking. But many workers came in fresh from the countryside, and it is fair to assume that the requirement of showing up at a determined hour and then having their work day further regulated by the clock was one of the real shocks of this new way of life.

Reactions and Reforms

Sheer necessity forced many early industrial workers to accept the factory work regimes, including definitions of timing, whatever their personal reactions. But there were two broad patterns of response, both indicating awareness of the burdens the new timing systems, combined with the machine-driven pace of work, placed upon the daily grind. The responses were in some ways contradictory, but many workers combined them with at least some degree of success.

Response number one was to use every informal method available to lighten the time pressure. Here was a common scenario: a machine breaks down, but rather than waiting patiently for the repair the workers involved start wandering off, chatting, circulating around the factory, simply reclaiming the opportunity to decide for themselves what to do with their time at work. Despite shop rules and rigorous foremen, some workers even found ways to sneak a nap on occasion.

Even more common was the impulse to get out of the factory altogether, when conditions allowed. Many skilled workers earned enough that they could periodically quit their jobs, even go back to the countryside for a bit, simply to regain control over their time and pace.

It is important not to exaggerate. Many workers were too poor to have any leeway. The foremen's discipline made it hard to reclaim control over time in the factory itself. Employers often exaggerated the ill-will of their charges, claiming that workers were so sly and lazy that they expended well under full effort. One French industrialist argued that the workers in his textile factory put in only 62 hours of real work a week, despite a theoretical requirement of 84.

Early industrialization involved a real battle over time, and employers, more powerful and more articulate, could not be relied upon to tell the truth about the efforts of their employees. But evidence of informal resistance is hard to ignore.

The second way to respond to the generalized imposition of clock time was to admit the validity of regulating work by the hour but to argue that the hours themselves should be reduced. Humanitarian members of the middle class might themselves respond to this argument, or even take a lead. Attention turned first to the burdens on child laborers and, sometimes, women. From the 1830s onward a variety of government-sponsored inquiries and private investigations revealed the deplorable conditions of many child workers, and also commented on the burdens of too much work on women. The response, very gradually, was to regulate work hours for these categories. A French child labor law in 1841 thus stipulated only eight hours for children under 12, 12 hours for children 12 to 14. Similar legislation was enacted virtually everywhere once factory industry began to spread. The laws were weak in many ways, poorly enforced, applicable only to big factories. But over time they improved and were gradually generalized to other work sites. And they always tried to respond to perceived problems by relying on hours as the basic unit of measurement.

Then in the second half of the century, spurred by growing labor movements, attention was also directed to limiting hours of work for adult men as well. Again, progress was slow and inconsistent, but by 1900 some categories of workers had won 10-hour days, and there were even some pioneering efforts, in particularly demanding occupations like coal mining, to bring the total down to eight. All of this confirmed the importance of clock time, but partially reshaped its impact.

Weekends

A further stride in this direction involved recasting the definition of the week, that old human timing creation, toward more clearly delineating a weekend. The movement had a somewhat complicated history. Traditionally, in many parts of Europe, urban workers had taken off on Mondays, sometimes called "holy Mondays", another sign of how irregular work time could be before industrialization. The idea was that if Sunday was for religion, workers needed another day for recreation, frequently including considerable drinking. The informality of holy Monday was incompatible with industrial efficiency, particularly since it often affected the quality of work on Tuesdays as well. So holy Monday was gradually phased out, though not without conflict.

The idea of a weekend, now defined as Saturday–Sunday, was not a literal response to the disappearance of tradition, but rather a more general effort to moderate the pace of factory work and give workers a bit more time of their own. Arrangements for Saturday afternoon off, as well as Sunday, began to pop up in some British factories in the 1840s, and the idea spread. The first explicit reference to the weekend, in English, occurred in 1879. By this point workers in other societies were pressing for what some of them called the "British week". The idea was that by allowing Saturday, or part of Saturday, for recreation, and then Sunday in principle for church and family, workers could show up Monday both rested and sober. An American textile company granted a five-day week in 1908, and other moves also took into account the growing minority of Jewish workers who needed Saturday off for religious purposes. Progress toward the two-day weekend was slow and halting – it still does not exist in some societies today – but the concept was clearly established as part of the first century of response to the advent of industrial time.

Continuing Tensions

The experience of new kinds of work time was surely most difficult for workers during the first generation or two of factory industry. This was when the exposure to clock time was most novel and unexpected. This was when employers were particularly likely to cheat. And this was when time burdens were unmodified by any serious gestures toward reform, even for child workers.

Many employers thought the acceptance of timed work improved with later generations of workers, partly perhaps because of some reforms but partly as well because workers simply became accustomed to the system. Some had worked under the clock as children, or had gone to schools where time was emphasized.

But tensions persisted even so, partly because, for many workers, long daily work requirements lasted well into the 20th century. Less skilled workers – British porters were working 14 hours a day in 1900 – and workers in small shops were particularly burdened. Even in many factories a (usually) quiet battle between employers and workers over control of work hours continued.

This was why, late in the 19th century, many factory workers were claiming that their control over work time was diminishing compared to looser conditions in early industrialization. It was becoming harder to goof off a bit, or even sneak a nap. The supervision of foremen was becoming more efficient. Machines themselves were getting bigger, requiring more constant attention: by 1900 a weaver was overseeing up to 16 mechanical looms, compared to the one to two looms of his counterpart six decades earlier. Added to this, in some industries where hours of work had been formally reduced, there were recurrent pressures to work overtime – a new concept, sometimes rewarded with additional pay, but sometimes simply imposed. Even though the notion of clock time was no longer seriously in dispute, its meaning and impact were still evolving.

Further Innovations

Two additional developments around the turn of the century offered further extensions of the application of the clock to work.

The first was the time clock. The first time clock was invented in 1888, by an American jeweler named Willard Bundy, though some similar devices may have predated his effort. Time clocks took advantage of the spread of electricity, and allowed a detailed record of entries and departures so that tardiness could be carefully noted and pay determined exactly by time spent on the job. For many workers, this was simply a new and more precise mechanism to enforce what shop rules had long sought. But the introduction of the time clock not only confirmed the association of clock and work, but extended it to additional occupations. Department store clerks, for example, or telephone operators had to check in just as factory workers did. White collar workers, a growing category, were increasingly pressed into modern time frame.

The second innovation was at least as important. From about 1900 onward, and headed by developments in the United States, a new profession of industrial engineers began to study the components of manufacturing work to determine how timing standards could be perfected. Appropriately called time and motion studies, this research carefully charted the movements of workers in order to generate more efficient norms. The goal was, as auto maker Henry Ford put it, to make each worker as machine-like as possible, Industrial engineers used stop watches, among other devices, to figure out the time requirements of each phase of the production process. The most famous of the new industrial engineers, Frederick Taylor, thought that workers were deliberately slowing things down, keeping their supervisors ignorant of how rapidly work could really be conducted. And his new systems, quickly extended to assembly-line production belts, were designed to take time control even further away from workers' hands.

As what was sometimes called "taylorization" spread to Europe (and soon to Russia), workers became widely aware of this further imposition on their experience of work. "Why should we be asked to work under a system that indicates every movement of our elbow", British typographers lamented in 1908. "In the old days they used to use a lash upon the slaves; in these days I suppose they would use an indicator." Workers had long since lost control over time at work. Now many of them were losing further control over the ways timing was defined and applied.

The introduction of the assembly line might be combined with important sweeteners. Henry Ford, an ardent proponent, also offered relatively high pay and, in 1926, introduced a five-day week. Work might play a slightly lesser role in daily life. But the new systems confirmed the basic industrial trend for work itself: the regulation of the clock, now including the minute and second hands of the stop watch, reigned supreme.

Travel Time

During the 19th century, beginning in Western society, transportation and time became increasingly linked, with major results in both policy and the human experience. The developments had been anticipated by new uses of clock time in the stage coaches of the 18th century, and in the growing importance of accurate time in navigation. But developments from 1820 onward carried the importance of accurate time to new levels – both for the transporters and for the people being transported. The results were probably not as significant for most people as the changes in the workplace, but they unquestionably supported the increasing organization of human life around carefully timed schedules. And, as with work, the involvement with more precise time was associated with a growing concern for speed and efficiency.

Innovation showed quickly in transatlantic shipping. Before the 1820s, ships between Europe and the United States had normally left when enough cargo had accumulated to make the venture worthwhile. Exporters and even human passengers – including would-be immigrants – simply had to wait around, sometimes for weeks, until the departure was announced. But by the 1820s, as volume steadily increased, a number of shipping companies switched to regular scheduling, listing the day and even the half day of departure, with at least a projection for the date of arrival. This shift was facilitated by the advent of more rapid sailing ships, including the famous *Yankee Clipper*, and by growing competition for the fastest-possible crossings. Then in the 1840s, when intercontinental steam shipping at last became feasible, the importance of scheduling increased still further.

The advent of shipping schedules contributed to a steady increase in the levels of traffic in goods and people. Business people and even tourists began to be able to schedule regular trips. The pattern would spread to shipping lanes in other parts of the world. The amount of patience needed to contemplate long-distance travel declined considerably – though frustration when schedules were not met might increase as well.

The assumption of regular scheduling applied almost immediately to the advent of train travel. But here the focus was not on a day or half day, but an hour and even a precise minute. Railway lines began to develop in the 1820s, initially in Britain and, from the 1830s onward, gained ground throughout the Western world and, soon, beyond. A new relationship with time was present from the outset.

There were several reasons for marriage between the clock and the train – signaled, as train stations began to be built, by a prominent clock for the benefit of passengers and rail personnel alike. Trains were far more expensive to operate than traditional means of land transportation, and it was important to take full use of their availability: scheduling departures at a set hour and minute would facilitate maximum use. There was no waiting around for a crowd to gather. Even more important was the issue of safety. Many routes operated with a single track, punctuated periodically by a doubling where trains, heading in opposite directions, could safely pass each other. Precise timing was essential to avoid accidents – and in fact, even with the best arrangements, calculations periodically broke down, prompting still further attention to time coordination.

The advent of the telegraph supplemented the improvements in timing. Information could be sent from station to station about times of departure and arrival, using clock time as the common currency for information.

As trains expanded in length, clocks and watches even served as communication devices among rail personnel. Knowing the precise minute of scheduled departure, a train conductor would prominently consult his watch to make sure passengers knew they must get on board; and the engineer, consulting his own watch, could assume that everything was in order for an on-time departure.

By the middle of the 19th century, the "railroad watch" became a vital component of train travel, designed to assure the greatest possible accuracy for engineers, switch yard controllers, and conductors. Rail companies increasingly sought to regulate the quality of the watch, or chronometer, and acceptable standards for accuracy in one period were periodically replaced by more demanding criteria. Various watchmaking firms, as in the United States, specialized in the manufacture of railway watches, and then might seek to translate their reputation for precision into sales to a wider public.

A new profession emerged, calling for detailed knowledge of timing requirements and timepieces. In the 1890s a leading watch inspector, Webb Ball, was hired by the Lake Shore and Michigan Southern Railway after a crash that resulted from poor timing, when two commuter trains collided. Ball's expertise in bringing this company's timing up to standard led to work on a number of other lines – ultimately, about half the rail companies in the United States – helping to promote more uniform criteria nationally.

By the early 20th century, detailed specifications defined the manufacture of railway watches, including a large, visible dial, with Arabic numerals; a stem at 12 o-clock; a minimum of 17 jewels in the watch movement; and a daily variation of no more than four seconds from watch to watch (or 30 seconds a week); plus protection against variance amid changing temperatures. Time, on the rails, was a serious business. Ultimately, the United States government also got in on the act, assuring time standards through the Naval Observatory.

The unprecedented attention to time as a vital element in the advent of train travel had several implications, beyond the involvement of train personnel themselves in the timing effort. Policy coordination was a vital issue, surprisingly difficult but easy enough to track. Impact on passengers is another consideration, inevitably harder to assess.

As railways companies worked on the issue of timing by the mid-19th century, they faced two related issues quite apart from demands on watch technology. Different units long operated on different time standards. The criteria for watches on one line might differ from those on another. This was a particular problem in the United States and Britain, where a wide variety of rail companies competed; in many European countries, government regulation from the outset reduced competitive differentials. In the United States, the American Railway Association met in 1887 to agree on a fairly standardized set of requirements, but not every company fell in line.

More pervasive were the tensions among different local time systems. That is, despite wide agreement that a day had 24 hours, of equal length, considerable disparities persisted in defining when the day began: as had long been the case, noon in one region might occur at a different time than noon in another. Variations were partly a matter of tradition, partly a reflection of the fact that the position of the sun differed depending on a region's longitude.

The result, for several decades, was a mess. Travelers had long needed coordination tables to figure out the timing standards of whatever locality they found themselves in; but now, with trains, they could pass through several different patterns in a half-day. For a while, some watchmakers tried to address the issue by manufacturing products with two dials, which could be set to different local systems.

By 1873 the United States featured a full 71 different local time patterns. European countries faced similar variety. German rail companies worked steadily to generalize the timing system of the largest regional city: thus Munich time was gradually imposed in southern Germany, Berlin time for the north. Partisans (in an age of growing nationalism) urged that these standards could serve as "great national clocks". Train companies routinely listed the schedule of a train run based on the time system of the city of origin, and this continued to differ greatly from the local time systems through which a train passed.

In Germany and elsewhere, moreover, a surprising amount of local resistance impeded standardization. Traditional time systems reflected not only local geography, but a certain amount of local identity as well, and there was real hostility to what was called the "aggression" of the big rail operators. In Britain, for example, the Scots and Irish were particularly hostile to efforts to impose "English" time. But the problems of variability were widespread. When the great transcontinental rail line was completed in the United States, with the symbolic ceremony at Promontory Point, Utah, in 1869, (at 12:45 PM local time), it was reported as 11:44 AM in San Francisco, 12:30 PM in nearby Virginia City, and 2:47 PM in Washington DC.

Ultimately, as long-distance train travel became more routine, the confusion proved simply unacceptable: train companies could not operate sensibly, and travelers themselves were needlessly confused. Connections among train lines were very difficult to determine. Britain made the first move toward coordination, defining rail time using the meridian of the Royal Observatory at Greenwich (near London) – referred to as Greenwich Mean Time. The coordination process began in 1840 but was completed, by a Royal Decree, only in 1880. France moved later, generalizing Paris time across the country in 1891, and Germany defined national time in 1893.

The problem was more difficult for the United States and Canada, simply because of their size and the different meridian positions that resulted. In the late 1860s a New York teacher, Charles Dowd, developed a proposal for a national time system. He initially thought in terms of a single "railway time" pattern, allowing localities to keep their own systems, with travelers therefore needing to juggle two timetables at any given point. But then he moved to the idea of major regional time zones. Backed by some of the big rail companies, a General Time Convention met in 1873, ratifying Dowd's work.

Still, nothing actually happened. More than a decade passed, filled with a great deal of advocacy by railway time officials. Finally, a new Convention was ratified in 1883, defining standards known as "Railroad and Telegraph Time". Five zones were defined, from the Atlantic to the Pacific (four for the continental United States), each varying from its nearest neighbor by one hour. At noon on November 18, 1883, the zones were established, with telegraph lines transmitting time signals to all the major cities. The agreement allowed localities to keep their own system if desired, providing conversion tables from the national grid. But, gradually but inexorably, the larger, nationally defined time zones came to predominate. The whole episode, in North America and elsewhere, was an interesting exercise in the tension between local preferences and new coordination demands, and the fact that it took a few decades to resolve is not really surprising.

What did railroad time mean for the population at large – even before the national decisions sorted out some of the difficulties? For many people, surely, already accustomed to carrying watches or looking at a public clock, the advent of precisely scheduled travel probably seemed nothing more than a logical extension. But there were stories of peasant villagers in Europe, suddenly confronted with a need to take a train trip, coming to the station hours before the train left simply because, having no timepieces of their own, they did not know how to organize more precisely.

Even for more sophisticated urbanites, train times added to the fact that people's days were increasingly scheduled around the pace of the clock, and not more spontaneous human rhythms. The impact was particularly great on that group of people that came to be known, by the 1840s, as commuters. In many cities – such as Boston, New York, Chicago in the United States – rail companies quickly set up spokes from urban centers into the emerging suburbs. To encourage use, they "commuted" tickets for regular travelers, offering a lower rate – hence the new terminology. As early as the 1840s, tens of thousands of commuters were coming into and out of center cities each workday, operating by the clock not only for their work time, but also now for their travel time as well.

Only a few professional categories managed to modify the rush. Most urban banks established later hours of operation, so that their suburban managers did not have to follow the timetables of their harried colleagues quite so precisely. Revealingly, the resultant schedules were known as "bankers' hours". But for most suburban workers, particularly until the advent of the automobile, the combination of travel and the job meant that a large portion of the waking day was rigorously defined by clock time.

And what of the impact on human patience? In principle, as with more precise ship times, the advent of clock-based schedules for land travel placed fewer demands on patience. Instead of waiting around for a stage coach to depart, when it finally filled and the driver was ready, the train traveler now knew when his conveyance would leave and what timing to expect thereafter. But there was of course a modern catch. Trains did not always leave on time, yet clock-based expectations, the belief one should be able to arrive exactly when planned, made delays seem more intolerable than ever before. Conceiving a full history of human patience is not an easy task, but it seems likely that the increasing conversion to calculating in terms of hours and even minutes made even minor disruptions harder to endure than before. Certainly, "making the trains run on time" – one of the boasts of Italy's fascist dictator, Mussolini, in the 1920s and 1930s, became a standard measurement of good management in a modern society.

Leisure and Recreation

One of the well-known results of industrialization, clearly taking shape in the cities of Western Europe and North America by the later 19th century, was a fairly rigorous division of the day between work and nonwork, or leisure (plus sleep, of course). As we have seen, preindustrial days had been less clearly divided. During working hours, except in periods of extreme pressure like harvest time, people had chatted, wandered around, even napped somewhat at random. Industrial work patterns explicitly sought to disrupt this pattern, though not always entirely successfully. In principle at least, work was for work; shop rules said no chatting, no singing, certainly no napping all under the eye of the foreman/supervisor.

As hours of work gradually declined a bit, this did raise the possibility of reserving part of the day not needed for outright rest, for other kinds of activities. Many advocates of the eight-hour work day explicitly thought in terms of a threefold division: a third for work, a third for sleep, a third for recreation. A group of immigrant workers in Worcester, Massachusetts, in the late 19th century, opposing municipal efforts to regulate what they did in the public parks, directly stated: they wanted "eight hours for what we will". The claim already recognized the structuring power of the clock, of course. The whole notion of the weekend, gradually gaining ground, also assumed the importance of giving workers and employees time to use as they wished.

But the push for freedom from work constraints also raised the question: to what extent, as daily leisure segments became better defined, would people be able to keep precise timing out of their day after work was over? To what extent would they want to? This section explores the ways in which new kinds of timing began to penetrate even nonwork life, either against people's wishes or with their connivance. There is no question that the clock was less burdensome in leisure hours than at work – but it gained a growing role.

Untimed Options and Their Limits

Several leisure patterns, either traditional or new, implicitly suggested a desire to avoid precise timing in the part of the day that was under more personal control. Many factory workers crowded into bars after work or at least on weekends, where time constraints might be minimized. In many continental European cities, simply taking a walk with the family, on a Sunday afternoon, was another popular and unstructured pastime.

There were newer options as well. The rise of baseball in the United States, from the 1840s onward (with the first professional teams forming in the 1860s), was an interesting example for players and spectators alike. Games had starting times, obviously, though they might not be too rigorous. But timing was not involved once the game got going, and – for those with enough afternoon leisure to show up – the result might be a refreshing contrast to the dominance of clocks in other aspects of life. Other rising sports were also untimed: cricket in Britain, gymnastics in Germany and other parts of central Europe.

Remnants of the festival tradition also persisted, though more in rural areas than urban. Here too, daylong feasts and entertainments followed no particular clock. But the festival was in trouble. Modern employers did not like festivals – they occurred too irregularly and took up too much time, leaving participants sluggish even when the event was over. Urban authorities also distrusted festivals, for they grouped crowds of people in ways that might turn disorderly. And, amid growing cities, some festival traditions were hard to maintain simply because familiar village associations were lost amid urban strangers.

So: a major form of untimed leisure was now in decline, which raised questions about what kinds of leisure options might possibly compensate. Add to this the attacks on traditional mixtures of recreation and work on the job, and there was a clear potential void.

One other component was vital: popular leisure, in the cities particularly, was under intense scrutiny during the 19th century by middle-class moralists and even some labor leaders eager to see workers use their time constructively. We have seen that the old prejudices against popular choices, the belief that ordinary people were too likely, left to their own devices, to use leisure time unproductively, gained further currency. Worries about drinking habits visibly increased. A good bit of police time, in early industrial cities, was devoted to trying to discipline popular recreational behavior, including in and around the taverns. Moralists did not rule the roost. Workers resisted, often successfully.

But the notion that leisure should be expressed in new ways might gain support as well, not just from the moralists but from urbanites themselves who wanted to take advantage of new opportunities, particularly as hours of work began to decline a bit. And, for various reasons, leisure innovation might well involve new uses of the clock, even though this meant that leisure might overlap with some aspects of work in unexpected domains.

Imposing the Clock

Precise time entered the leisure world in several ways. Regulations were one source, which is where middle-class moralism came in. Growing concern about popular drinking habits and intemperance led to some efforts to ban alcohol entirely. Far more common, however, was the simpler device of regulating tavern hours.

Thus the Wine and Beer House Act in Britain, in 1869, allowed localities to set fixed times for activities in the classic "pub". Enforcement varied, and so did the specifics of legislation. During World War I opening hours were rigorously restricted to noon to 2:30, and 6:30 to 9:30, with strict policing. In most periods, however, there was greater variety. But the imposition of the clock supported the classic closing call of the publican: "Time, gentlemen, please."

The second intrusion of time simply involved the growing dependence on scheduled transportation, particularly after around 1860. Leisure-goers had to keep some eye on the clock so they would not miss the last commuter train or subway. Vacation travel, again usually by train, required some careful planning. Organized jaunts to the seaside, popular in industrial Britain by 1900 (with workers often solemnly sitting in the sand in their Sunday best, because they did not know how to swim), required knowledge of train schedules and access to a watch or clock to keep time.

The clock's third entry into the new world of mass leisure involved the growing need to operate popular theaters, and then movie houses, on a fairly strict timetable. Once gas lighting and then electricity made their entrance, opportunities to stage several shows a day became irresistible – an obvious contrast to Shakespeare's day. American vaudeville and British music halls thus ran two or three discrete showings. But to make this manageable for the performers, and to make sure audiences assembled coherently, clock-based schedules became essential. Admittedly, they might permit a few minutes' leeway, but flexibility was limited.

All this means that the 19th century was also the century in which the middle classes and increasingly the working classes began to apply the clock to the domain of leisure. Given the novelty involved, the results could seem almost as stressful as the experience of time at work. Thus, an article in the London *Times* lamented, referring to the need to plan recreational schedules precisely: "It is work, it is tiring work, it involves precise attention to time. And all the anxieties and irritations of that responsibility."

But the most extensive intrusion of time into play centered on the substantial transformation of athletics, away from casual games and toward clock discipline. Three changes overlapped.

First, as with urban theater, schedules became essential to coordinate the arrival of crowds, often from some distance. As professional sports emerged, with soccer football and baseball venues, game times had to be established clearly. Of course, individuals could still choose to come late – a particular option in the relatively untimed baseball setting. But the clock did intrude.

Second, a major movement in the rise of modern as opposed to more traditional sports was the growing effort to regularize rules, to allow fair competition among different localities and, often, to impose some discipline on what had often become rowdy affairs. Part of the establishment of rules, in turn, was agreement on the duration of many games, and divisions within this duration. Football (soccer), notoriously, came to operate by the clock, with extra minutes carefully doled out to account for disruptions during the 90 minutes of regular play. American football, though far less continuous, established clear timing for quarters and halves, and also the length of "time-outs", as well as the game itself. And the same would apply to the even newer sport of basketball. Boxing too, one of the three most popular spectator sports in the United States in 1900, was segmented into timed rounds – another use for a bell.

As noted before, this aspect of timing remained a variable. Several sports resisted almost entirely, though as with baseball or tennis this could provoke recurrent anxiety, in a time-conscious society, that games were too unpredictable (and sometimes clearly too long). Crowds did not necessarily stay for whole games – they could modify their clock by personal convenience. And informal games, even in the regulated sports, might have no set duration at all – as in playground basketball games or some versions of bare-knuckle boxing. With all this, however, the marriage of many modern sports to the clock became increasingly widely important.

And, finally, within many sports – even some of the less-timed ones – came a concomitant modern fascination with speed, and a desire to time and record speed, typically aided by the equally modern stopwatch. This interest was well-established in horse racing, where times among other things would affect sale prices. But now it extended far more widely.

Track and field events began to be more formally organized from the 1860s onward, with the formation of groups like the Amateur Athletic Club in Britain (1866) or the Amateur Athletic Union in the United States (1887). One of the several functions of groups of this sort became recording times in the sport and publicizing champion performances. With the establishment of the modern Olympic Games in 1896 and then (in track) the International Amateur Athletic Federation (1913), speed times went international: one of the prime functions of the new IAAF was to approve world records. The fascination with timing affected the interested public and individual athletes alike, shaping how they judged the quality of effort.

Similar impulses soon applied to swimming and the new sports of bicycle and then automobile racing. Sports where speed records were not fully applicable, like American football, were nevertheless increasingly affected by a desire to check the running times of key performers. The impulse was widespread.

Many historians, and many ordinary fans, have long viewed sports as an alternative to the pressures of daily life: a chance to get excited, to share with crowds, to enjoy distinctive settings like the green fields of baseball or football. And these perceptions are accurate to a point. But modern sports also reflected and promoted wider values that were, and are, shared with the workplace, and the adherence to clock time is at the top of this list. Sports brought time home to many people who might otherwise have preferred to ignore it. The nature of sports, and their role in public life, changed significantly in the process.

Sleep

One of the fundamental alterations in modern perceptions of time involved sleep. It is clear that major changes occurred in sleep by the 1920s – probably mainly from the 1880s onward, though details of the process are not entirely clear.

The big shift was the gradual disappearance of the old notion of two phases of sleep, with a period of wakefulness in between. The concept was still referred to in the late 19th century, though infrequently, and it seems to have disappeared entirely by the 1900s. In its place, in principle, came a single uninterrupted period of sleep, to which notions of clock time could be more clearly applied.

Causes of the change include the now familiar shifts in the nature of work and, above all, the advent of widespread use of artificial lighting. For many people, even before electricity, the need to be up early in the morning in response to the factory whistle, and then stay at work for 12 hours or more, made it difficult to afford the indulgence of more loosely timed sleep. Going to bed at a particular time, and even more obviously waking up at a particular time, now substituted for the older pattern, with clocks, including alarm clocks, increasingly measuring the process.

New forms of lighting were vital, reducing the natural time distinction between day and night. Quite apart from nights on the town, it became increasingly tempting to stay up later and then use the clock to determine when it was time to go to bed. By the 20th century, both popular and medical discussions of sleep were almost entirely framed in terms of the hours that should be devoted, with no assumption of any mid-phase interruption unless sleep problems were involved.

Concern about timed sleep began also to apply to children. The first studies of children's sleep emerged in the late 19th century, as part of the rise of pediatric research. Interestingly, initial experts – the first research paper came from France, in 1897 – recommended about 75 minutes more sleep for youngsters than is fashionable today. But a consistent theme, again from the late 19th century onward, was that modern kids were getting too little sleep. The *British Medical Journal* blamed the stress and bustle of everyday life, trolley cars, and gas street lights. A 1905 study bemoaned the fact that "this is a sleepless age and more and more… we are turning night into day". By the early 20th century conscientious parents could consult scientific advice on how best to schedule their children's sleep. A new wave of childrearing advice, by various medical popularizers and government pamphlets, offered careful instruction on how many hours of sleep were appropriate for each age.

Similar patterns emerged in the United States. Childrearing advice in the mid-19th century ignored sleep issues, and indeed children like adults seemed often to sleep in irregular bursts. But here too significant shifts set in by 1900, including regular lists of the hours of sleep desirable at each year of age. Expertise contributed to this new interest, but so did middle-class innovations like putting infants in cribs, in their own room, rather than having them sleep next to parents or with other siblings; and a decline in the use of live-in maids, which put care for sleepy children more directly in parents' hands. It became more important than ever to try to get children, from infancy onward, on a clock-based sleep schedule. Early 20th-century experts (overoptimistically) urged: "It is important to start the baby on a regular schedule when he's tiny; most hospitals will help you work out such a schedule and train the new baby to it for a few days before he goes home." For slightly older children, not only health but success in school depended on fixed times for sleep, often including a scheduled quiet time before bed. Regular timing was crucial – as in life in general: "nothing should interfere with the bedtime habits – neither evening shows nor guests".

Overall changes in the timing of sleep, and growing reliance on the clock for measurement, generated a fairly consistent array of concerns, and not just for the young. In fact, people often were getting less sleep than experts recommended, thanks to the various pressures and distractions of modern life. But the level of anxiety was also interesting, as many adults, from the early 20th century onward, recurrently tracked their actual sleep times and compared them to the latest scientific standards. New uses of time, and new levels of worry, were often linked. In many cases, deep belief in the importance of timed sleep schedules was built in from childhood and, while often ignored in practice, could still fuel anxiety.

Socializing Children

The importance of instilling a time sense in children followed obviously from the emphasis on time in general – particularly in the workplace. And here emphasis developed well before the emergence of efforts to schedule their sleep.

An American historian, not a big fan of the quality of public education in the 19th century, once claimed that, if they accomplished nothing else, schools at least taught children clock time – which would then loom large in their lives as adults. And there was no question that many educators and moralists, including some fervent Protestants, looked to the schools to correct what they saw as the laziness and loitering of the working classes. Thus some British charity schools, shortly before 1800, were praised for requiring students "to rise betimes and observe Hours with great Punctuality". Another proponent of the importance of "regularity" noted that a new Sunday school for girls stipulated that each student should be in place by nine in the morning, and again at 1:30 in the afternoon, "or she shall lose her place the next Sunday, and walk last". In a Methodist Sunday school, the first rule was: "I am to be present at the School…a few minutes before half past nine o'clock."

But it must be remembered that the interest in preaching precise time to large numbers of children – as opposed to relatively elite school attendees – was a new one. Injunctions against childish idleness go way back, but linking this to the clock was not an established tradition except in a few urban settings, and then mainly for the more prosperous families. Most children had simply been coordinated by the orders of a relevant adult; they did not need an independent time sense and, when not working, there is every indication that they were given a lot of latitude for unstructured play. Furthermore, even in the 19th century schools caught on only gradually, particularly in rural districts where many children dipped in and out depending on work needs back home; they may not have learned very much about time (or anything else). Serious enforcement of attendance began only in the early 20th century. Many schools, in addition, mixed children of various ages, teaching lessons in different parts of the schoolroom with no particular specification of how much time should be devoted to a given subject. Some classroom monitors, children themselves, may not have known how to tell time at all. So we cannot rush too quickly in bringing children into the picture.

As in other aspects of the history of time, we know that, ultimately, clocks did begin to regulate children's lives and that this socialization was an important part of the larger imposition of timing in the long 19th century. But knowing the trend, and developing a better understanding of the precise process, are not the same.

Even the fabled school bell had a complicated evolution. Many schools had bells to announce the beginning of classes – but they were often rung by individual teachers, not necessarily according to a rigorous schedule. Bells became more common in the later 19th century. In addition to school bells, individual teachers bought hand bells to signal when the class needed to shift to another subject, or to announce beginnings and endings of recess and lunch periods.

Schools were not the only source of socialization, of course: for many working-class children, there were the factory whistles; and for almost all children, there were parents. As with schools, the evidence about parental time discipline, during much of the 19th century, is somewhat vague. Lots of childrearing manuals moralized about the importance of teaching hard work, but they usually did not tie this to the clock. Thus one American domestic manual, in the 1840s, urged that children learn they had "no right to waste a single hour" – but clocks were not explicitly brought into the equation. Punctuality received some attention: the same manual noted that "unpunctual" students usually did badly in school. And there was some discussion of the importance of early rising – but again, not always by the clock: in winter, one advisor urged children to get up as soon as there was enough light to get dressed. Punctual bedtimes were also recommended. But there was less explicit insistence on the importance of learning to go by the clock, during the day itself. Far more attention was paid to qualities like obedience, piety, and other more traditional virtues.

Other evidence, however, suggests that many older children were fully aware of time and clocks by the middle of the century, even if we cannot be sure of the precise nature of their training, or the age at which a time-sense was expected. Thus in the wildly popular novel, *Little Women,* references to several clocks in the house, and their chiming during the night, crop up frequently. And a children's club, mirrored after fashionable adult secret societies, met routinely at 7 PM every Saturday.

But probably the most famous literary indication of children's familiarity with time – sufficient familiarity to know that an exaggeration was meant to be funny – came in Lewis Carroll's famous *Alice in Wonderland,* written in 1865. One of the more striking characters in this fanciful story was the White Rabbit, usually pictured anxiously consulting his pocket watch and muttering frantically, "oh dear, oh dear, I shall be late."

Finally, whatever the variety of experience in the early 19th century, schools unquestionably turned to more systematic attention to time by the 1870s. It was at this juncture that more rigorous and systematic use of bells began to dot the school day. By this point, larger, consolidated schools and more systematic curricula took some of the decision power away from individual teachers. It became more important to designate class periods, carefully timed, so that students could pass from one subject, and its teacher, to another. Many schools (at least after the primary grades) were shuttling students from one subject and teacher to another, including shop and "home economics" classes plus music periods, by 1900, and this would have been impossible without rigorous timing. Installing electrical systems in the schools also began to permit building-wide buzzers, replacing more harmonious but less effective bells. Gradually, timing, and its sound effects, were beginning to become a fundamental feature of the school day. Teachers, as well as students, were meant to comply: thus a 1885 Ohio directive allowed instructors to keep misbehaving students for 15 minutes after classes ended – but no more. Clocks were for everyone.

Schools themselves were teaching time directly at the primary level, both in reading and in arithmetic classes (arithmetic itself, not a common subject in Western education before the 19th century, gained ground steadily). A first-grade primer in the *McGuffey* series – the most widely used in the United States – turned to time directly in the 1876 edition, in a reading lesson that referred to a "fast train" that would get a father to the station by 8:30 – but after good children's bedtime. A second-grade reader offered explicit discussions on how to read 10 o'clock. Another item pictured a clock directly, with a presumably catchy jingle: "Can you tell me what time it is?/ Look at the clock and then tell me./Long hand for minutes and short for hours./How many hours in the day? How many minutes in the hour?"

And the lesson ended with a pithy statement about the regulatory qualities of modern time: "Our clock at school tells us many things. It tells us when to work and when to play."

For many people in Western society, the long 19th century was clearly the point in which thinking, and behaving, in terms of hours and minutes became a daily reality. The change centered on work, but extended widely beyond.

Variety persisted. Rural time continued to differ from urban time, though the percentage of the population in the countryside was dropping fast and interactions between rural producers and urban agents were increasing. Many women, working primarily in the home at least after marriage, had less pressure to work by the clock than men did, except for the challenge of getting the family ready for jobs or school. Women's domestic work might be arduous, extending through a long day, but it was less clock-based at this point. Even the very time-conscious American domestic manual written by Catherine Beecher in the 1840 – urging the "the right apportionment of time to different pursuits" – admitted that precise scheduling by the hour was impossible for a wife and mother; she proposed assigning tasks by the day, with Monday for laundry, Tuesday shopping and so on. Time-consciousness was involved here, but not the clock. Jokes about female punctuality – their stereotypic problems in getting ready to go out on time – may have reflected a real gender distinction, into the 20th century. (One misogynistic example: "The punctual husband is a lonely man.") More durably, university students formed another group, though a small one, less committed to clock time. Formal classes operated by a schedule, to be sure, though there were some margins for error: in Dutch universities a custom developed of beginning a class 15 minutes after the official start time, a tradition that has now been eliminated. But outside of class, time structure was less apparent in university life – except to the extent that young people had already been socialized to the clock, in earlier schooling.

Conversion to the clock also promoted the intensification of prejudice. Even in the early modern period we have seen the generalized complaints about the working class as lazy and undisciplined. These complaints persisted, now applied more specifically to punctuality. Some groups were singled out explicitly for their more traditional approach to time. Immigrant groups might encounter this bias in the United States. And it was early in the 20th century that references to "African time" or "Colored People's Time" emerged in the United States, in this case persisting for many decades thereafter. African Americans, in turn, would handle this cultural claim in various ways – on into the 21st century, as President Barack Obama once noted.

Acceptance of the industrial clock was a struggle, even as some eager publicists claimed to embrace it with pleasure. Schedules were imposed on many people. They were painfully built into the socialization of children. Acceptance may have eased over the course of the century, but even in 1900 tensions persisted, though inventions like the weekend presumably helped a bit. The experience of time had moved further away from nature, and adjustments could still be difficult.

One important indicator may at least suggest some degree of accommodation, however hard won: whereas at the outset of the long 19th century fines were widely imposed to try to compel an often reluctant labor force into acceptance of the clock, the emphasis on fining receded considerably by 1900. Of course, people who were habitually late risked getting fired; the element of compulsion did not disappear. It was in the 1920s that a German office manager even installed a steam jet in the toilets, to make sure that staff (primarily females) did not spend more than two and a half minutes sitting down. And time clocks helped link work and pay without the burden of fines. Still, it is probable that more urban people by 1920 simply knew that they normally had to get to work on time, without further enforcement, than had been true a hundred years before (though whether they still resented this or not is another matter). Industrial time may have become a bit less burdensome given a century of adjustment.

There is also evidence, at least from the 1840s onward, of the growing importance of clock time in family life, quite apart from the socialization of children. Several accounts highlight a new cause for marital quarrels, when one spouse (most commonly the husband) fails to get home at the time promised. Family punctuality became virtually a moral responsibility.

And there is another ambiguity, important to consider in evaluating the modern concept of time. Developments in the 19th century encourage us to think of time as an imposition or constraint, and so it surely was for many workers, school children and the like. But modern time-consciousness has another angle, in facilitating the realization that some enjoyable experiences pass much too quickly. While occasional references to time passing "swiftly" go back to Shakespeare in the English language, it was only from about 1800 that the common phrase, "time flies", began to catch on widely. Here was another, sometimes rueful, way to express the growing awareness of time.

One final point is a bit removed from the time-consciousness of ordinary people in Western society, but intrudes into Western cultural history by the end of the 19th century. Several scientific developments, apart from the world of clocks and watches, had important implications for a larger sense of time. From the 1830s onward, work in geology extended the understanding of the immense duration of time in the experience of the Earth. Then, with Darwin's *Origin of the Species*, the sense of the extent of evolutionary time added to this impact. The new ideas provoked dispute around religious issues: many Christian traditions had assumed that God created man and the universe only a few thousand years ago. But even with the disputes (that continue in Islam as well as Christianity), the new discoveries might promote an interesting sense of the limits of the human experience compared to the vastness of Earth time. A sense of "cosmic time" began to emerge, at least among relevant scholars, focused on the short span of the human experience compared to the age of the Earth: if the existing of our planet is taken as one day, human beings have been around for only four seconds, a possibly humbling thought.

In the final decades of the 19th century, in another scientific sector, new research – centered particularly in the growing interest in electricity and the speed of light – began to question the absolute qualities of clock time. Henri Poincaré, a French scholar working in the Bureau des longitudes, pondered the question of the standardization of local times around the world, building in insights from a Dutch theorist, Hendrik Lorentz. The result was a somewhat complicated theory that posited the idea of relativity in time, when clocks were in motion, but preserved the more classic belief that there was a true "real time" as well. These ideas would feed into the more elaborate research on relativity and time, and particularly the ideas of Albert Einstein, in the 20th century.

As with geology and evolution, the partial reconceptualization of time involved complicated scientific research, and had little connection with the time frames in which most people operated. It has led to speculation, however, that there were some wider shifts taking place in Western culture at the turn of the century, ready to call some of the older certainties about time into question.

Further Reading

The classic article on time and work discipline is Thompson, E. P. "Time, Work-Discipline, and Industrial Capitalism." *Past & Present* 38, no. 1 (December 1, 1967): 56–97.

See also Glennie, Paul, and Nigel Thrift. "Reworking E. P. Thompson's 'Time, Work-Discipline and Industrial Capitalism'." *Time & Society* 5, no. 3 (October 1, 1996): 275–99.

Ingold, Tim. "Work, Time and Industry." *Time & Society* 4, no. 1 (February 1, 1995): 5–28.

Postone, Moishe. *Time, Labor, and Social Domination: A Reinterpretation of Marx's Critical Theory*. (Cambridge; New York: Cambridge University Press, 1996).

Stearns, Peter N. *Lives of Labor: Work in Maturing Industrial Society*. (New York: Holmes & Meier Pub, 1975).

Westenholz, Ann. "Identity, Times and Work." *Time & Society* 15, no. 1 (March 1, 2006): 33–55.

On railroads:

Schivelbusch, Wolfgang. *The Railway Journey: The Industrialization of Time and Space in the Nineteenth Century*. (Oakland: University of California Press, 2014).

On time and leisure:

Bailey, Peter. "'A Mingled Mass of Perfectly Legitimate Pleasures': The Victorian Middle Class and the Problem of Leisure." *Victorian Studies* 21, no. 1 (1977): 7–28.

Cunningham, Hugh. *Time, Work and Leisure: Life Changes in England since 1700.* (Manchester: Manchester University Press, 2016).

Rosenzweig, Roy. *Eight Hours for What We Will: Workers and Leisure in an Industrial City, 1870–1920.* (Cambridge: Cambridge University Press, 1985).

On sports:

Gorn, Elliott J. *A Brief History of American Sports.* (Urbana: University of Illinois Press, 2004).

Guttmann, Allen. *Games and Empires.* (New York: Columbia University Press, 1996).

Polley, Martin. *Sports History: A Practical Guide.* (Basingstoke; New York: Palgrave, 2007).

On sleep:

Ekirch, A. Roger. *At Day's Close: Night in Times Past.* (London: W. W. Norton & Company, 2006).

Stearns, Peter N., Perrin Rowland, and Lori Giarnella. "Children's Sleep: Sketching Historical Change." *Journal of Social History* 30, no. 2 (December 1, 1996): 345–66.

Wolf-Meyer, Matthew J. *The Slumbering Masses: Sleep, Medicine, and Modern American Life.* (Minneapolis: University Of Minnesota Press, 2016).

On manners:

Kasson, John F. *Rudeness and Civility: Manners in 19th Century Urban America.* (New York: Hill & Wang Pub, 1990).

On reconsiderations about time at the turn of the century:

Kern, Stephen. *The Culture of Time and Space, 1880–1918.* 2nd edition. (Cambridge, MA: Harvard University Press, 2003).

Rudwick, Martin J. S. *Worlds Before Adam: The Reconstruction of Geohistory in the Age of Reform.* (Chicago, IL: University of Chicago Press, 2010).

6 Time in the World During the Long 19th Century

In 1884 the President of the United States, Chester Arthur, convened an International Meridian conference in Washington, to discuss choosing "a meridian to be employed as a common zero of longitude and standard of time reckoning around the world". Delegates from 41 nations attended, almost entirely from Europe, North America, and Latin America, though Japan was there, and the British colony of India was represented. Votes were not unanimous, but the conference fully agreed "that it is desirable to adopt a single prime meridian for all nations, in place of the multiplicity of initial meridians which now exist". Almost all attendees also concurred on "the adoption of a universal day for all purposes for which it may be found convenient" (which did not preclude use of local time as well), the day to begin at midnight and to contain 24 hours.

In its own quiet way these decisions constituted a high point of late 19th-century globalization, while also illustrating the extent to which it was largely under Western control. It made adoption of rational standards seem easy (though the mixed voting showed that coordination in fact might be difficult). But the felt need to standardize time at this official level, and to insist on a single rational pattern, was not the same as the ability to require conversion away from more traditional measures of time. Here, the late 19th century was a time of struggle and resistance, in many parts of the world.

The history of time in the world as a whole after about 1800 revolves around several fundamental questions. To what extent would people be willing to accept increasingly common time standards and devices? To what extent would they be compelled to do so, in an age of new work regulations and imperialism? And how widely and effectively could resistance be expressed, and on what grounds?

Two developments predominated, and they ultimately intersected. First, Western traders and imperialists increasingly interacted with the time systems established in other major regions of the world. The result was a substantial effort to impose the concepts and mechanisms now widely established in Europe and most of North America – not only clock time, but also calendars. The project bore many resemblances to the impositions occurring in the West itself, and for many of the same reasons, particularly around real or imagined work efficiencies. But imperialist impositions often involved assumptions of local inferiorities that could be even fiercer than the prejudices directed against subcultures in the West, with strong overtones of endemic idleness and inferiority. On the other hand, regions, even under imperialist control, sometimes had a stronger basis for resistance to what were demonstrably foreign time systems, than lower-class groups had maintained within the West.

There was, however, no single formula for regional reactions. Much depended on prior time systems, which had varied widely. The invitation to comparative assessment is obvious. Many of the comparisons will amplify distinctions that show up in other facets of world history in the period, amid varying patterns of imitation and resistance.

The second main development – also largely in the hands of Western organizers – involved the efforts, particularly in the final decades of the century – to coordinate time on a more fully global basis. These projects built strongly on the systems that had already responded to the needs of the railroads, but beyond individual continents. The result was the system of time zones that is still in play today, though some subsequent modifications have occurred. But coordination did not always come easily, for many people saw these arrangements as another part of an effort to undermine local time identities with foreign impositions.

Both aspects of this essentially global history of time reflect the fact (at least, according to many relevant historians) that the process we call globalization was taking serious shape, at least by the second half of the 19th century, thanks to new technologies and trade levels, but also a variety of new international arrangements and even the first phases of the internationalization of sports. The need to reconsider time not just in one society, but potentially in all societies, both reflected the increasing contacts that were generating globalization, and contributed to the process in turn.

The first part of this particular story, the development of global arrangements, is relatively straightforward. But the second part, the interaction of regional players with Western-dominated definitions of time, is far more complex.

Time Zones

The second half of the 19th century saw a number of international agreements (almost always organized by Western nations and then generalized as widely as possible), designed to coordinate international communications and other contacts. A postal agreement made it possible to send letters from one country to another, based on acceptance of the national stamps of origin. An International Telegraph Union formed to coordinate policies in this domain, with clear implications concerning the need for shared time standards. During these decades as well, leading scientists were increasingly meeting on an international basis, arranging to exchange meteorological and health data and to coordinate statistical techniques. Here was a context that could easily be applied to discussions of time.

Fortunately, a good bit of spade work, on debating standards of time across geographical space, had already been applied to the knotty problem of the railroads, with North Americans often taking a lead in conceptualizing wide regional zones that could reduce the problem of purely local time definitions yet take adequate account of the unfortunate fact that the visibility of the sun varied greatly by longitude. Even before rails, relevant preliminary definitions had followed from the urgency of calculating longitude: this was the issue that had prompted the British Royal Observatory to adopt Greenwich Mean Time as a reference point in 1675.

There was precedent, in other words, for thinking about regional clusters and about the possibility of taking one particular meridian as the regional standard in turn. Add to this, by the second half of the 19th century, the wider enthusiasm for thinking about ordering the world as a whole and the growing need, at least for some businesspeople and imperial administrators, to cut through idiosyncratic detail, and a context for further definition was clearly established. In the eyes of leading reformers – and "time reform" was in fact the term they used – it was now possible to create a uniform and entirely rational system for the measurement of hourly time. As international business and communications links steadily expanded, there was considerable momentum in their favor.

Though many Westerners in fact, including many political leaders, continued to have trouble getting rid of traditional local definitions of time, even in the face of standardized railway schedules, reform leaders believed they could approach the problem of time on a global scale with complete rationality, independent of political or identity concerns. Time in their view – by which they meant contemporary European definitions of time – was open to orderly and universalizing arrangements that would apply to the world standards that were increasingly taken for granted in Western cities and factories.

Various precedents accumulated. In 1868 the British colony of New Zealand became probably the first country in the world to adopt a standard time that would be observed throughout the colony, basing it on a longitude 11.30 hours east of Greenwich. The North American railroad agreement, in 1883, on five zones (Intercolonial for Atlantic Canada, then Eastern, Central, Mountain, and Pacific) was another major step. There were some holdouts from the new system – Detroit, near the borders of two zones, insisted on its own local time until 1900; and the national government mandated acceptance only by the Standard Time Act of 1918. But the principle was clear well before this.

Various proposals had already circulated at the international level. An Italian mathematician proposed 24 hourly time zones around the world, based on the meridian of Rome, and urged the importance of this kind of coordination both for astronomy and telegraphy. But his ideas won little attention in his own lifetime. A Canadian, Stanford Fleming, proposed a different time zone system in 1879, and began to win considerable attention at international conferences.

Then, in 1884, the scientists and diplomats at the Washington Prime Meridian Conference urged a global system of time zones based on the Greenwich Meridian, though it was agreed that they did not have authority to mandate this arrangement. They did, however, insist on international agreement that the day begin at midnight – an obvious challenge to localities that had long used sunrise, noon, or sunset – though they did offer a bow to tradition: the new measures were not to "interfere with the use of local or standard time where desirable".

Setting up a system and winning acceptance were two different matters, of course. Even European countries bickered for a while. France – England's long-time global rival – balked at the idea of Greenwich as the baseline, and continued to use Paris until 1912. But most countries had accepted the time zone system or a fairly clear variant by the late 1920s. Russia, which as we have seen was often quite ready to participate in wider arrangements of time, first identified a "Moscow zone" as early as 1880. Ultimately some regions, because of meridian position or tradition, opted for a half-hour deviation from the standard zones – as with Newfoundland in Canada, Iran and Afghanistan, and parts of Australia. And a few were slow to sign on to the basic system altogether: Nepal was last, in 1954.

But for most of the world, a regional relationship to the global time zone system was accepted, at least in principle, fairly quickly. The whole process was aided not only by the growing importance of global contacts and transactions but by the fact that the decisions for many regions – such as Africa – were taken by Western colonial regimes and not by indigenous agreement. Still, the system did seem logical – it survives today – and most new nations when they won independence confirmed their acceptance, though with occasional modifications to take special national needs into account.

All of this, however, could seem pretty remote from ordinary life, since most people were not engaged in frequent international travel or communication. Only the insistence on starting the day at midnight could be a jolt – if it was actually applied to daily activity. The real interaction between regional cultures and global standards came not through debates over Greenwich Mean Time, but through efforts by governments, businesses, and missionaries to force or cajole an explicit change away from local habits – in the interest of standardization, or authority, or efficiency, or orderly morality, or some combination of these values. This is where the globalization of time became truly complex, even though a broadly common process was involved. And this is where a real test of wills could take shape – or, in some cases, not.

Regional Encounters

Calendars

The long 19th century was not a period of major global adjustment for yearly calendars. Western imperial regimes brought the Gregorian calendar with them, but local systems usually coexisted and remained more important for most people – and for the crucial scheduling of religious events. When the United States purchased Alaska from Russia, the calendar shifted from Julian to Gregorian, with the usual results that several days were skipped over, during October 1867. At the same time the International Date Line was moved west, so that it separated Alaska and Russia rather than Alaska and Canada.

The most interesting decision came from Japan, eager after 1868 to fit into global patterns, particularly to coordinate business dealings. The government officially adopted the Gregorian system in 1872, adjusting to it in the following year. This was no small shift, as a host of personal recalculations were required, but it did establish a newly global orientation. (For comparison, think of the personal and institutional adjustments required to move from traditional measuring systems to the metric pattern – adjustments so great that the parochial United States has still not tried it.) Most Japanese long resisted using the BC/AD dating system for historical time, basing chronologies from key royal periods instead, but even this habit would gradually yield to the Western tradition during the 20th century. Korea followed a roughly similar pattern, opting for the Gregorian in 1895 but using a prior royal dynasty rather than the Western (Christian) BC/AD distinction.

But for most regions outside Europe and the Americas, where conversions had occurred earlier, the big decisions about calendar time awaited the years after World War I, when in some respects at least globalization shifted into higher gear. During the long 19th century, in countries like Russia, maintaining the existing calendar system was an important statement of identity and religious commitment.

One other point is worth nothing, for it crops up again today: as time was reconsidered in the 19th century, a number of reformers even in the West thought that the Gregorian calendar itself could be greatly improved upon. George Eastman, the founder of the Kodak photography company, and Eliasbeth Achelis, an American activist who became known as the "calendar lady", thought that a brand new start was essential for a scientific, rational age. They urged, quite simply, 13 months of 28 days each – a notion first put forward by a French technocrat. Nothing came of these ideas at the time, for Western leaders were more concerned about winning fuller global assent to the system they already had and were not interested in taking the trouble to change their own habits. To be sure, internationalists in the early 20th century continued to call for a new global system, which would be religiously and culturally neutral as well as uniform. The League of Nations, in the 1920s, correspondingly pursued the project of a new "world calendar" with some vigor, but there was no outcome and interest faded.

Clock Time: Japan Again

As in so many domains in the later 19th century, Japan proved to be something of a special case, eager to adapt to international systems and Western influences without losing an independent identity. The Japanese also had an edge thanks to their earlier adjustments to mechanical clocks. The fact that Japan also pressed forward with railway and shipping development formed another context in which new timing systems could be considered.

One vital symptom of a willingness to change was the quick popularity of imported Western clocks and watches (brought in from Europe and the United States alike), once the nation ended its substantial isolation after the 1850s. The personal commitment was unusual. Most Japanese, well into the 20th century, struck to traditional consumer forms for the most part, as in choices of food, furnishings and dress; but watches were an instantaneous hit, viewed as an essential means of adjusting to the modern world or, as the reformers put it, to "civilization and enlightenment". Again, the fact that the nation had already generated interest in clocks and watches, though tuned to a different hourly system and available mainly to a wealthy elite, helps explain the interest in further steps.

But real change was involved, and not just a dissemination of preexisting systems. Traditional Japanese clocks, for example, had uniformly maintained only one hand – there was no attention to finer divisions such as minutes, and as we have seen even the hours were seasonally variable. The key policy shift, which supported the new enthusiasm for clock imports, was the decision to standardize the length of the hour. And this shift was part of the larger acceptance of the new calendar system – solar rather than lunar – by Imperial Edict in 1872–3. The government, insisting now on the fixed hour, framed the change vigorously:

> [I]n order to modernize our country and reform the ancient customs with the goal of moving forward into the realm of civilization as a people, it is of utmost urgency that we hereby rectify the calendar law.…The present system whereby one day is divided into one hundred hours, defined in duration by the length of day and night, creates great inconvenience in its general application to all enterprises. …The (new) system, which has helped our country's government in all spheres since the establishment of international ties with other lands, must be adopted.…We hereby order that the solar calendar be published throughout the land, using an equal number of hours to divide day and night, thereby reforming the time system, not only in order thereby to put right the calendar law, but to promote the enlightenment of the people, as well.
>
> Whereas, until now, the system of hours has been comprised of a rendering into twelve hours varying in duration in accordance with the length of day and night, we hereby newly establish a system of time comprised of twenty-four hours equal in duration for both night and day, dividing the period of time lasting from the hours of the rat until the hour of the horse into twelve hours, and naming this period "forenoon", and dividing the period of time lasting from the hour of the horse until the hour of the rat into twelve hours, and naming this period "afternoon".

This was, of course, a very dramatic decision, and the idea of condemning the past (in the area of time – not in every domain) in the name of progress was meant to highlight a change that otherwise might have seemed a concession to the new needs of coordinating with foreign businessmen. The approach certainly reflected the authority of the Meiji government itself. And it also recalled other situations in which a ruling group had sought to change the regulation of time as a demonstration of power.

For most Japanese, the shift meant, of course, that existing clocks and watches became immediately useless, setting the stage for a long period of imports. Actual change was, as is so often the case where time is concerned, more gradual. Many people continued to use older time units, shunning minutes and regular hours, and even today older terms (such as "horse" for noon) continue to be used. Rural Japanese, particularly, had no particular reason to change their habits, even though the calendar change was periodically thrust upon them when they needed an official document. On the other hand, reformers were eager to spread information about the new time system, including how to read a modern clock; one pamphlet to this effect, by an educational leader, quickly sold hundreds of thousands of copies.

The government itself promoted change by beginning to construct public clock towers, deliberately designed "to educate people to regard this time as their own". The clock faces themselves were initially imported from the West, but in 1875 four Tokyo artisans got together with a plan to build four towers by themselves– "with the aim of providing a public good that will last for all eternity". A number of private shops also began to feature public clocks, headed by the flagship store of the Minatoya Meat Company.

Scattered data show the clear progress of at least urban interest in clock time. In 1862, before the great reforms, only 1,400 Western clocks and watches were sold. But by the 1880s 18 import houses were handling merchandise from several countries, with the Swiss in the clear lead. By this point over 700,000 clocks had been purchased from abroad. But by this point as well, Japanese businessmen were also commissioning cheaper watches made in small shops in Japan itself. Local industry was also stimulated by the need for repair facilities, even for Western visitors after a long journey across the Pacific. Only in 1888 was a major Japanese production operation established, using machine tools for the first time and quickly generating 1,000 grandfather clocks per year – though a prejudice for foreign models would linger for many years.

Clocks long remained more important than watches, meaning that many urban Japanese still depended on public installations. Again in 1887, 30,000 schools, over 10,000 town halls, 4,500 post offices, 4,800 police stations and numerous hospitals and banks had public clocks. Watches spread as well, but cost remained a factor; they were, however, a popular graduation gift in the upper classes and the military. One gold watch was proudly given to a policeman who solved a particularly gruesome murder in 1898. Cheap German alarm clocks also won a market share. By the early 20th century, purchase of slender wrist watches for women began to catch on, with Swiss imports again in the lead.

The main point, of course, was the combination of access and interest, and here again the evidence points to fairly rapid change. In 1877 only one in 30 Japanese households had a modern timepiece of any sort. By 1897 the figure was one in three, meaning that most urban families had one. By 1907, clocks were to be found in over 70% of all households, which shows that they had effectively conquered many rural areas as well. These figures were higher than the rates in Europe at the same point, and higher than many regions in the United States. Watches, again, spread more slowly – only 10% of the population had one by 1900, though they were now essential for government officials and the military. But rates of ownership expanded rapidly during the first decades of the new century.

The speed of change in Japan during the Meiji or Reform era after 1868 has often been noted. Obviously, a transformation in time sense forms a vital part of the overall list. With amazingly little overt pushback, a complex but well-established traditional time system was shifted to the Western-style clock. To be sure, the fact that traditional systems might persist, particularly in the countryside, helps explain the process of adjustment, but even here change entered within mere decades. Arguably, no society – even in the West – ever made this kind of conversion so rapidly.

A number of components contribute to the explanation. Foreign example, and the availability of relevant foreign products at not too steep a price, were vital factors – though again the Japanese did not convert to foreign styles in other areas despite similar potential inducements. Prior experience with relevant technology – mechanical clocks – surely helped, though of course they had supported the older patterns. The authority of the government, and its willingness to frame change in terms of civilization itself, clearly played a crucial role, carried out not only by decree but by active sponsorship of public clocks. A habit of considerable obedience and the fact that time was presented as a national priority and not a foreign imposition (despite the reliance on imports) were crucial here. Consumer example – the desirability of keeping up with the neighbors – and the real-time demands of a rapidly industrializing society complete the picture, along with the eager implementation of clock time in the rapidly expanding public education system. Surely, some Japanese must have lamented the need to change, in what is a very personal aspect of life; but their concerns never found widespread expression.

The key factors involved were intriguing but, again, also unusual. The speed and smoothness of Japanese reaction to Western time systems simply were not matched in other regions. More commonly, a combination of confusion and resistance slowed the acceptance of clock time – though change did occur.

The Ottoman Empire: Imperial and Turkish Experience

The situation of the Ottoman Empire, in the long 19th century, was measurably different from that of Japan, and reactions to new systems of time reflected this fact. Though technically independent, the rulers of the Empire had much more trouble managing outside interference than their Japanese counterparts did. A major effort at reform in the middle decades of the century failed to produce the kind of transformation that the Japanese managed just a bit later. Time, in the Ottoman Empire, was also more intimately linked to religious beliefs and identities than was the case in Japan. All of this added up to a context in which outside pressures, generating real change, also promoted new levels of complexity and confusion.

No basic calendar reform seemed necessary or possible. The Islamic calendar, and its lunar basis, provided an alternative to the Gregorian approach, and Middle Eastern and North African societies had already made practical accommodations to the solar year in the calendars they kept for economic activities.

Mechanical clocks were another matter: they could not be resisted entirely, and members of the Ottoman elite had in fact been importing some fancy devices even before the 19th century. This process continued after 1800, and many mosques even began to incorporate mechanical clocks, often grandfather clocks, as part of their notation for times of worship. Initially however these imports were combined with traditional daily time patterns, including the division of night and day into 12-hour segments, with hours varying in length depending on the season and clocks reset every day to make sure the day ended, at 12 o'clock, with sunset. No major rethinking was required in this first stage of change.

By the middle of the 19th century, however, this time system was clearly beginning to interfere with international business dealings and with government efficiency, because of the continual adjustments required. During the Crimean War coordination with British and French allies (joined with the Ottomans in an effort to contain Russian expansion) was difficult because of differences in time sense. Even earlier, the launch of the overall Ottoman reform process, in 1839, raised inevitable timing issues. Crucial to the reforms was an expansion of government services: more and better courts of law, telegraph and postal offices, rail and ferry operations. But these activities almost required agreement on formal hours of operation and also pushed for a greater sense of punctuality among the Ottoman public. The rise of the telegraph itself imposed change: coordination with telegraph offices around the world was virtually impossible without some agreed-upon time frame.

Some of these issues were shared by the West itself in the same decades. The problem of having various local times, depending on longitude, was not unique to the Ottoman Empire, and it would be sorted out only gradually even in Europe and the United States. But the presence of the Islamic system of flexible lengths of hours, and its association with religious identity, formed a more acute problem, not only complicating time coordination but supporting potential resistance as well.

Ottoman reformers, and the European businessmen and officials who gained increasing influence in the Empire, pressed for a rationalization of time, based on the Western clock sense. This was part of a larger standardization effort and the deployment of new levels of political authority. But for many ordinary people, even in the cities, and for many decades, this new definition of clock time coexisted with the cherished local religious systems. People operated on two (or more) definitions of the day or hour, depending on what activity was involved. Transportation schedules often simultaneously listed Western and Turkish times, with no small amount of confusion resulting.

The government, however (even as it pulled back from some reform activities that were viewed as dangerous to the sultan's authority), pressed ahead with further attempts at coordination. It wanted active obedience to what it insisted was a more rational order. To this end, the reigning sultan, Abdulhamid II, decreed in 1901 that every city in the empire build a clock tower in his name. The principles of these clocks was clearly stated, as the sultan ordered "the construction of a clock tower in a suitable position which will show western time and automatically chime the hours". Clock towers had become one of the means that the ruling elite sought to demonstrate the government's power throughout the provinces.

Public clocks had in fact been gaining ground since the early 19th century, but gradually and at first with no sweeping policy claims. Some of the early towers were located near mosques, occasionally using an old Christian tower in a structure that had been converted to a mosque. While bell chimes seemed foreign and possibly religiously offensive, many Muslims came to accept the early public clocks as a modern version of the old Muslim service of timekeeper: a new way to help schedule the call to prayer. A poem in 1884 thus celebrated a clock tower as a "perfect timekeeper house in this neighborhood": "When a great 'bell clock', like no other in our land, Came from Europe it gave everyone great joy. No longer is there need to carry a pocket watch, With its sound, of time the world comes to know."

The new and more systematic program of public clocks did not change the equation entirely. The government made sure that the new towers, for example, were not as tall as the nearest minaret. And the sultan himself referred to the potential utility of the clocks for the calls to prayer.

But the new program was intended to encourage a new and more secular sense of time, not primarily to blend with traditional patterns. Most of the towers were built near other government buildings in a central city square. This aimed at maximum public visibility but also served to regulate the work schedules of government employees themselves. Thus a tower in the city of Adana "chimed every half hour, and public officials began and ended work" according to its signals.

The new public clocks could indeed be used, of course, to note Muslim prayer times, and some commentators tried to claim that the government was calling for prayer: "Such a huge masterpiece that none can compare….For day and night, the tower announces the time of prayer."

But there were two problems with this effort to blend old and new. First, the bells used in the public towers sounded suspiciously like Christian church bells – which Islam had, from the outset, deliberately shunned. (Even when Christian worship was tolerated in Muslim-ruled societies, as was often the case, the use of bells had been forbidden.) In some cases the government compounded this association by extending tolerance to additional Christian groups. All of this set the stage for authorizing some Christian churches themselves to build bell towers, which angered Muslim leaders directly. And, second, Muslims already had very effective means for the call to prayer. Many Muslims argued that the new chimes were set to the "time of a foreign land", and ignored or resisted them accordingly. In some cases, in fact, new clocks were sponsored by European governments in an attempt to curry Ottoman favor; this might easily enhance local suspicion.

What was going on here, at the turn of the century, was essentially a battle between the old commitment to organize the day according to religious needs, and a new desire not just to insert government authority but to arrange the day primarily in terms of times for work. This was a tension that had emerged in Europe as well, but earlier and with more gradual adjustments. During the latter days of the Ottoman Empire, the confrontation was stark.

As had occurred in other areas, political enthusiasm for public clocks led, by the early 20th century, to explicit urban rivalry. Cities competed with each other to build towers, sometimes taking funds away from sanitation programs for this purpose. Local governments, in other words, were placing a priority on their role in converting public consciousness. The chimes of public clocks were deliberately intended to reach into residential areas and even private homes, forcing attention to this new organization of time in terms of the 24-hour day.

All of this set the stage for an even more decisive policy move, once the Ottoman Empire itself collapsed in the wake of World War I. A new, republican government in Turkey, under the nationalist reformer Kemal Ataturk, pressed the authority of the state further than the Ottomans had dared, eager to bring the new nation more fully into the modern world. The new government in 1926 passed a measure, called the Adoption of Western Time and the Gregorian Calendar Act. The reform regime was less involved with compromises with Christian minorities or explicit efforts to please foreign powers – issues that had colored reactions to Ottoman timing efforts. But the authorities were bent on completing the process of conversion to Western time that had evolved more gradually during the 19th century itself. True to form, another spate of public clock construction, often now linked to more modern architectural forms, resulted.

And some traditions were displaced quite clearly. The role of timekeeper houses, near mosques, was fairly steadily undermined: the new clock time simply made them unnecessary. Turkish life, and not just in official circles, was increasingly organized according to the new system, including explicit attempts to instill the value of punctuality.

But, in Turkey and in many other largely Muslim regions of the former Ottoman Empire, the result was not a pure triumph of one system over another, just as clock towers themselves had sometimes been assimilated for religious purposes. Muslim prayer times continued to define some of the public space, along with the symbols and sounds of the newer system. Calls to prayer coexisted with the chimes of public clocks, and electronic amplification gave them greater range than before. Many people were able – and continue to be able today – to operate according to two time systems and the sounds associated with each.

All of this represents a somewhat different pattern from what had occurred in the West, or in Japan, though the process of change embraced many standard elements and, in the Ottoman case, some direct foreign involvement. The role of the state certainly can be compared to patterns in Japan, with important differences in chronology (suggesting that the process of change was more difficult in the Ottoman case) and thoroughness. As in Japan, and in fact in the West, the evolution was guided by regional leaders, not primarily through explicit outside imposition. The needs for change emanated from Ottoman (and then Turkish republican) political calculations combined with some of the seemingly inevitable needs for coordination in the modern world. The result, again, was an intriguing compromise with remnants of an older system, producing a distinctive modern urban soundscape even as secular timing needs gained ascendancy.

The Ottoman Empire: The Arab Experience

Arab culture, in key parts of the Ottoman Empire such as present-day Lebanon, obviously shared in many of the results of official policy – before the Empire collapsed – but also interacted with other influences in the complex dance around pressures to change the sense of time. Ottoman government policy had less role in predominantly Arab regions; and there was no outright imperialist intervention. Nevertheless, a number of prods for change emerged within the region, forming yet another case study in the globalization of time – or lack thereof.

Lebanon, and its key urban hub of Beirut, already harbored a number of patterns of time. Large Arab-Christian populations, mainly Orthodox, followed a Julian calendar, while a substantial Jewish minority had its own scheme, and the Muslims stuck with the lunar model. Calls to prayer by the *muezzin* defined key aspects of the day, with the day itself beginning at sundown. But another time system, called "Frankish" or European time, had the day beginning at noon. A public clock tower, built in 1897, competed with the clock towers of the several Christian churches. Multiplicity, in other words, was the predominant context.

Pressures to rationalize or to follow the new global time zones system had little influence. Well into the 1920s local almanacs had to spell out what international time was all about, and what major cities of the world had adopted it, which means that it had, as yet, had scant impact amid the mixture of local systems.

Christian missionary schools, from the late 19th century, pushed the idea of Western clock time. One principal in a new Syrian Protestant college complained about the lack of a clock, and so ordered one from the United States. Newspapers discussed time reform and, by the 1900s, featured a number of articles on famous clocks around the world. Several newspapers published schedules of the various time systems (prayer times, steamboat times, and so on), though different papers offered different specifics – which prompted many reader complaints. Some Arab reformers, eager to see their society adjust more fully to the modern, industrial world, began to write about systems of time in terms of "backwardness" and "progress".

Discussions of time were enhanced by a small crisis in Islamic time calculations in 1910. Disputes arose over the precise timing of the end of the holy month of Ramadan (this was not unprecedented in Islam, as different regions might proclaim slightly different times for moon sightings). This fed into a larger discussion of whether it was legitimate to use the telegraph to convey Ramadan sightings from one area to the next. All of this contributed to a larger debate about how and whether to use at least some elements of Western time calculations.

Clocks and watches spread to growing sectors of the urban population. Swiss imports were advertised as coming from "the biggest factories in Europe". Local manufacturers claimed they had mastered modern watchmaking machinery – one owner bragged that he had learned the techniques in Australia – while there were also many complaints about shoddy standards. Some successful local manufacturers won awards for their work, from groups like the Protestant school: one winner was cited as having been "among those who spent their time…on useful inventions for the nation". A good bit of reform literature, in Arabic, also revolved around a combination of traditional hostility to idleness and a more specific injunction to manage time wisely – adopted among other things from Western publicists like Samuel Smiles. Widely circulated translations of *Self-Help,* from 1874 onward, highlighted new virtues like "punctuality". "Lost time is gone forever." Picking up on these themes, Arab writers now referred frequently to concepts like "killing time", lamenting how many people wasted hours in cafes, oblivious of the value of time itself. Lecturers urged the importance of change, on topics like "man's time, from the benefit of saving it and taking interest in it, from the vice of wasting and losing it". "Your happiness lies in saving your time."

These were lessons that took hold for many. A young man in Beirut in 1909 decided to go on a stroll pretending to be a European, wearing Western-style clothes and, of course, sporting a watch. But then he realized that it was very un-European to walk around aimlessly, so he went back home to use his time studying something that would help his career. Here was a new time-consciousness put to work.

Other writers, picking up on the idea of clock time, also invoked the rather different and more traditional notion of time in the sense of fate over a lifetime, which might have diluted the message somewhat, since fate was seen, by Muslims, as resting in God's hands, not in the scope of dutifully punctual individuals. But Islam itself could be invoked by the reformers: "God changes not what is in a people, until they change what is in themselves" – a verse from the Quran. Overall, however, purely Western obsession with time was modified, even among the advocates for change, with a more Islamic context where time was viewed in a very distinctive way. Writers picked up on European criticisms of local lack of punctuality, but mixed these with a larger concern that the region maintain its independence from European control and purely foreign standards. More efficient time management was essential, but it could be turned against the threat of colonialism.

From all this, there are several key points. In the first place, regions like Lebanon did not face the same kind of government pressure that surfaced in Japan and the Turkish portion of the Ottoman Empire. This makes the fervent discussion of time, and the growing fascination with watches, all the more intriguing. Western influence and criticism played a role in sparking new interest, as did the presence of missionary institutions. But the local sense of a need for change, amid awareness of greater Western strength, played a greater role. This would contrast, as we will see, with much greater tension over time changes in areas, like India, that were under direct colonial control. But the new interest was itself not entirely Western, but something of a blend with older concepts of time.

And with all this it is vital to remember that Beirut, well into the 20th century, did not change its overall willingness to juggle a number of different systems of daily time. Here, the absence of firm government direction or imperialist imposition showed through clearly. Most people could figure out their own combinations, including of course the new interest in more efficient time management. Only under new French colonial control, after World War I, was there increasing pressure to regularize Lebanon's position in the global pattern of time zones and to push for more uniform standards for days and hours. Even here, as in the other Ottoman realms, growing acceptance of clock time would continue to combine with the rhythms of the daily calls to prayer.

India

Patterns of time in India during the long 19th century, or in the African case discussed in the following section, featured one stark difference from the evolutions in Japan or the Ottoman realms: the presence of direct colonial controls was a crucial variable. As a result, though they shared many components with other regions, issues around time standards presented more opportunities for abrupt and insensitive imposition, based on beliefs about home-country superiority and pressing labor needs; but also more opportunities for bitter resistance. Over many decades, some basic outcomes were similar to those in other areas: time concepts were changed and to an extent standardized. But the process was distinctively complex and disruptive.

For India's British rulers were absolutely convinced of the moral as well as technical superiority of their time systems, and saw no reason to compromise with local patterns. Imposition of a new time sense was a way to demonstrate colonial authority – here, in ways similar to what governments did in other regions, though in this case with the additional foreign flavor. But it was also essential in order to run modern operations smoothly – including the train service that the British began to organize extensively from the 1850s onward; and to discipline what was widely seen as traditional inefficiency and lack of moral control. And while Christian missionary efforts in India won only slight success, missionary support for new time arrangements added to government initiatives and enhanced the moralistic component.

Time issues in India also turned out to be directly affected by the global time zone movement, to the extent that it encouraged British officials to press for standardization in the region. This added a rigidity to interactions with local populations, after the mid-1880s, that proved particularly problematic.

India of course had a rich time history of its own, including a great deal of scientific and mathematical research. Astronomers had calculated days beginning at sunrise and had mapped the subcontinent to determine solar positions in different longitudes, seeking to provide some basic standards. However, as was common in most places, actual time was kept on a regional basis, with each local kingdom setting its own parameters while referring to a Hindu calendar, with its combination of lunar and solar features, for religious festivals. We have noted the practice, well into the 18th century, of royal sponsorship of giant sundials to help determine local time accurately.

Some scholars have argued that Hindu concepts, emphasizing the unimportance of events in this world as opposed to timeless patterns of reincarnation and vast cycles in the universe as a whole, discouraged certain kinds of interest in time, compared to other cultures. But this often disparaging assessment should not overshadow the extent to which India resembled many agricultural societies in the extremely local view of time and the loose coupling of daily time with ordinary activities.

The arrival of British overlords did not have an immediate impact. But as early as 1802 a British official, the chief astronomer for the British East India Company, established a time standard for the subcontinent, called Madras time, relating it to the Greenwich Meridian back in Britain. A tradition was established of firing a canon at 8 PM outside the Madras observatory, to indicate that all was well with this new Indian standard time. While this time pattern was used, a few decades later, to regulate common train schedules across the subcontinent, it had little impact on other aspects of Indian life – except to establish a precedent that the British had both the right and duty to set times patterns for their new colony. Ultimately this would encourage Indian astronomers to work to expand their own observatories and participate more actively in discussions of time policy, but the main results of this activity awaited the 20th century and national independence.

Interest in addressing Indian time traditions gained intensity after the middle of the 19th century. There was a certain irony here: protests during the 1850s convinced the British that they must let up in efforts to reform local habits in many respects, for example in the area of family and social structures. Queen Victoria proclaimed, in 1858 after major riots the previous year, "a public equality between Europeans and Natives", professing "no desire to impose European convictions on any of our subjects". And indeed the colonial government did back off in key areas, for example in leaving the caste system essentially untouched.

But time, it turned out, was an exception, too important to leave alone. Some historians have argued that the new concern about time reform was in fact designed to reassert British authority after the riots had damaged its image. Certainly, growing interest – British and Indian alike – in developing industrial factories inevitably spilled over into considerations of time. Early factories in India had often operated rather informally, with workers drifting in "after the gates open" and going off again "when shadows begin to darken the workplace". This was a rhythm, of course, which had greeted new factories in many places, but it involved a level of daily variation and unpredictability that was increasingly unacceptable to the authorities. An official

Factory Commission, in 1891, lamented that "the Indian operatives do not work against their customs or their natural inclinations or their traditional habits".

The colonial administration itself began to move more aggressively into the time arena in the late 1850s, by building – a familiar step – a large number of public clocks, again with the goals of emphasizing British supremacy and promoting the virtues of punctuality on the presumably irresponsible urban masses (see Figure 6.1). Many new clocks were placed in public markets, which were often chaotic mixtures of shops and stalls over which the novel visibility of Western clock time might seem

Figure 6.1 Built in 1873, this tower shows some of the complexities of time and imperialism. The tower was designed by British architects to resemble the famous Big Ben tower in London. But it was largely paid for by a local Indian tycoon and dedicated to his mother who, as a religious Jainist, was not supposed to eat dinner before sundown; her son wanted a more precise way for her to anticipate dinner time. Picture credit: Marina Pissarova / Alamy Stock Photo

to impose a new kind of rationality. In a few cases, more traditional buildings were pulled down to create space for a clock tower, making the power dynamics – Britain over India's past – quite explicit.

Some clock towers were designed with religious goals in mind as well. One in an Indian Muslim community, arranged by an Anglo-Indian architect, took the form of a minaret with the clock at its heart. The notion was that local Muslims might ultimately be enticed into a church if they got used to clocks in a more familiar setting. Not surprisingly, locals were hardly fooled. One commentator wrote of this particular clock tower: "This Britishness was perceived in a like manner by both the colonisers and a larger number of Indians who interacted with them." Nevertheless, by 1908 at least 13 independent clock towers had been built on the subcontinent, and a far larger number of public clocks were located in government offices, city markets, schools, libraries, and churches – and, of course, railway stations.

In all this, tensions inevitably persisted between the colonial and business leaders, who thought in terms of rationality, progress, and efficiency, and ordinary Indians for whom standardized time – as opposed to local or unregulated time – was identified as "British time". Protests against modern time could be easily wrapped up in larger anti-colonial efforts, and it is often unclear whether the problem was mainly time, or lack of independence, or more local pride – or, most probably, some combination.

In 1881 a British regional administrator extended Madras time into the government offices of Bombay (now Mumbai), affecting the work schedules of all government employees – mostly Indians – as well as those who did business with them. The administrator, one James Fergusson, had himself missed a train because he could not figure out the mixture of local times involved – this gave his campaign a personal and emotional component that proved particularly problematic. The result was a confusing battle between people who accepted the new schedules, even though they really altered the rhythms of city life, and those who resisted. Some clocks in the city were now set to Madras time, but others were not, adding obviously to the confusion. As a local paper put it, commenting on the clashing systems within the same city: "Consequently great inconvenience is felt by a large number of people." It became vital not just to know two systems, but to figure out which individual was following which system. "Suppose I submit and keep Madras time, how am I to know that Jones and Smith do the same", an English-language paper noted.

But the fight continued for a while. When the University of Bombay asked for funding to illuminate its own, quite imposing, public clock, one group responded by offering money on condition the clock show Madras time, but another, opposed to the colonial administrator, insisting that the money would be given only to a clock that showed Bombay time. In this case the government ultimately folded and went back to Bombay time.

Then in the years after 1899 various colonial officials began calling for a common time zone for all of India, in keeping with the new international time zone system, so that an Indian zone would have a clear relationship to the global Greenwich system. Several scientific associations back in Britain were pressing for this coordination, noting in typical language that a uniform zone for India would finally abolish "the present barbarous arrangement, unworthy of a country pretending to civilization, by which every place keeps its own time". A time standard was decreed in 1906, but with flexibility for municipalities to opt out. But leading nationalist figures protested in public meetings, noting that there had been no consultation of public opinion and that the new standard was "artificial". "We are asked to forget our natural time, the same that we have been familiar with from times immemorial, and adopt the new standard." A variety of demonstrations similarly insisted that the new system was "against nature". Furthermore, massive labor strikes ensued, in which workers insisted that the new time would allow employers to extend working hours and would also disrupt religious observances based on solar time. Again, employers had to promise to return to local time systems, like Bombay time – and in fact a fully national standard was not adopted until the middle of the 20th century, after independence. Time, clearly, was a question of urban and possibly national identity.

Finally, in at least one case a popular riot, protesting various actions of the colonial government and sparked by a major disease outbreak, targeted the public market clock directly. Hindu and Muslim crowds combined. The market clock, that symbol of colonial control, was shot up as part of a larger series of riots. Historians have speculated that, in addition to disrupting the public market, the action carried a larger political and social meaning – possibly even targeting the clock as an instrument of wider human alienation.

For several decades, then, even as a Greenwich-linked time system was used in parts of the country, a number of regional variants persisted. Indeed, what one Indian paper called the "battle of the clocks" extended through the first four decades of the 20th century, with different localities accepting or rejecting the pleas for a national time standard. It is also worth noting that the national time system that was proposed, and gradually extended, was five and a half hours different from Greenwich Mean Time. Though a number of experts had proposed two time zones for the huge country, railroad officials and others preferred a single zone, and the half-hour adjustment was designed to limit regional hardships in relation to positions of the sun. This aspect of the compromise – a single national system despite a range of longitudes and therefore a wide range of daily experiences for the growing number of Indians who did live by the clock– survived into the establishment of a full national standard in 1950 and persists today.

Something in the combination of Indian culture, resentment against colonial control, plus British insistence on the importance of modern time, made time and clocks an unusually contested issue in modern Indian history. Despite references to "natural time" in the struggles with colonial authorities, many Indians simply did not accept the Western notion that one system of time was or should be absolute. The contrast with Japan and even the Ottomans is marked. An obvious conclusion is that it is far easier to get people – however gradually – to change their time habits if the authority involved is not perceived as foreign. The equation was particularly combustible in India given British confidence that their time was the only valid standard amid the official global move toward rationalization, and despite imperial pledges that local habits would be left alone. But there was also a particular quality in the time traditions nurtured in Indian culture that made new time standards particularly menacing.

Certainly, there is considerable evidence that the contests over time in colonial India had some lasting effects. Of course clocks gained ground. And a national time zone was ultimately agreed to after independence. Change did occur, even under British rule – as with the many, often ambitious Indian civil servants who kept within reach of clock time.

But time remained a prickly political issue, even after independence. India remains particularly eager to question some international time conventions, in the interests of reducing their foreign taint.

And many Indians, even as they adjust to clock time, continue to insist on a certain degree of flexibility, even in the present day – an ongoing result perhaps of the belief that there were no absolutes involved. Go back to the story that launched the first chapter of this book, of the Indian immigrant to the Netherlands who noted the contrast between Indian and Dutch definitions of adequate punctuality. Examine claims – open to careful verification; remember that a great deal of prejudice still attaches to beliefs about punctuality or lack thereof –that Indian factory workers remain unusually casual about time, showing up late and often leaving early and (some experts contend) jacking up labor costs considerably in the process.

Some of the tensions around time in modern Indian history are absolutely standard, reflecting the huge strain that clock time placed on popular assumptions about the ways life and work should be organized and the tendency of employers to combine insistence on time with attempts to extend effort. But there is little doubt that, if only because of the colonial variable, the long association between change in time with foreign impositions, adjustments in India were different – slower and arguably less complete – than was true not just in the West, but in places like Japan. Modern time creates some common pressures across cultural and political boundaries, but it does not generate a one-size-fits-all history.

Sub-Saharan Africa

Time issues in colonial Africa – beginning in the 19th century but extending well into the 20th – have their own distinctive contours – though with some clear overlaps with the Indian experience. Happily also, they have been widely studied, at least for some key regions. One vital caveat however: sub-Saharan Africa embraces many different regions and cultures, and many potentially important cases have yet to be explored. We already know that there was considerable variety in the African time experience, particularly because of very different precolonial patterns, though there are some dominant themes.

Pressures for change in the sense and use of time began well before the imperialist surge swallowed virtually the whole subcontinent. Foreign-owned businesses – factories, mines, even sugar plantations – were one source of change, similar in broad outline to the roles new businesses played in other societies in the 19th century. Government pressures would supplement, though on balance official policies or proclamations were probably less important in the African case than in Japan or India. Colonial governments played a vital role in attempting to support other efforts to enforce Western time, particularly through harsh punishments including jailing and whippings. But more of the rule-setting came from employers. Furthermore, for many regions, the role of missionaries was particularly prominent, in a polytheistic region where Christian (and also Muslim) religious figures became increasingly active.

Clocks certainly commanded a prominent place in many of the newly established Protestant churches. In 1818 the Anglican missionary board sent a clock to a new church in a rural area of South Africa, adorned with many decorative wooden figures. This particular venture misfired. Local Africans thought the figures represented evil spirits, and the clock had to be dismantled to demonstrate good faith. But other imported clocks continued to populate churches, and also missionary schools, throughout the century and beyond. Further, missionaries were more broadly convinced that imposing Western time systems constituted a vital means of improving conditions in Africa. In 1884 an American missionary observed, "Our natives will not be anything if they do not feel the propriety and necessity of forming habits of industry…as their easy-going ways do not furnish favourable soil for the Gospel." Another chimed in, resolving to make Africans around mission stations "feel as much as possible the value of time, labour and skill". Convinced that the Western time system was both rational and moral, and that local time habits were deplorable, missionary efforts generated steady pressure to replace traditional time systems. While Protestants took the lead in preaching the new concept of time, Catholics could join in as well. One agent in East Africa thus urged the need "to secure a modification of the African mentality, and to bring them to realize their duty as civilized men".

Colonial authorities certainly agreed, arguing that the best way to bring Africans into the modern world was to press them into new forms of employment, rigorous time schedules included. One British official in East Africa noted quite simply, "We have got to find legalized methods and force the native to work: I hope we may rely on the government to meet the case." Colonial administrations were as convinced as the missionaries were that Africans suffered from a "culture of idleness", though they tended to rely more on coercion and economic pressure than on religion to turn the tide. In 1917 the British governor of Kenya, again convinced of the need to make Africans "useful citizens", said quite simply that "the best means of doing so is to induce him [the African] to work for a period of his life for the European". By the 1850s and 1860s many colonial cities were making new efforts to convey clock time and using bells to denote crucial hours in the day.

But it was primarily at work that new time pressures were imposed most directly, particularly in the cities and mines but also in certain kinds of agricultural work, such as raising and processing sugar cane. European employers wanted to be able to set clear hours for the working day (often longer than Africans were accustomed to), with clock-timed breaks for lunch; they also often expected night work, for example during the peak of the sugar cane harvest, which also went against most African tradition. Further, European employers were as convinced as their religious and political colleagues that Africans were by nature lazy and undisciplined, incapable of voluntarily adhering to efficient time practices. By the early 1900s, if not before, the term "kafir time" began to be applied to local time habits – kafir (though of Muslim origin) becoming the popular derogatory European term for black Africans.

Much of this, of course, is a familiar framework by this point, given what was already happening in the West and in other parts of the world. What gave the

African experience a particularly complicated and often troubling twist was the extent to which the new time systems violated a wide range of traditional beliefs and practices concerning time. Far more than clock time was involved in many cases, which helps explain why tensions over time were so acute and durable through the 19th century.

For, prior to the arrival of Europeans, many Africans had maintained time systems that were quite close to nature. Their days began shortly after sunrise and ended shortly before sunset. Natural reminders, particularly for daybreak, were widely used – various calls and sounds from animals and birds. In some regions, there was no concept of a week (and hence no Sundays). Months were calculated by the moon, which made them roughly 28 days long (meaning that there were 13 months in a year), and languages like Zulu in fact used the same word for month and moon. In some cases important religious ceremonies were attached to the moon's waning stage, when the night was dark and it was believed that people should abstain both from work and from pleasure. Seasonal variations during the year also constituted important time divisions in the African calendar sense, in contrast to contemporary European belief that the year could simply be chopped up into discrete months with little attention to seasonal designations. Finally many crucial ceremonies were essentially untimed, based for example on the rituals that surrounded death in a family – very important but, in terms of time, quite unpredictable. In combination, these various patterns could and did feed the European belief that "kafir time" was both different and quite unacceptable.

All of this raised a host of potential clashes with the beliefs and felt needs of white employers. One planter, who clearly believed himself to be fairly enlightened, put the matter this way, in 1846:

> I am not one of those who, when they arrive in a settlement, because they see a number of blacks infer that they have a right to the labour of these persons, and that if they will not work for them they are set down as lazy scoundrels. But I am one of those who maintain that if a number of Kaffirs come voluntarily and offer their service, and accept service at a given rate of money wages, and for a specified time, that I have a right to the services of these persons, until the time expires.

Repeated tensions developed in the Natal region of what is now South Africa – taken by the British as a colony in 1843 – where the Zulu people concentrated. The month was a perennial issue. Europeans assumed that local workers should work through the Western definition of a month, often several days longer than what was assumed in Zulu lunar tradition. The workers themselves, expecting pay by the phase of the moon, were understandably outraged when their employers tried to insist that wages were not due until the 30th or 31st day. Another 1846 report:

> This afternoon, because I would not pay a kafir whose month is up on the last day of the month, I was abused like a thief. He shook his stick at me, and was so violent that if I could have got assistance, I would have sent him to (jail).

Another European employer commented in 1855, in a classic statement of mutual intolerance:

> The month of service (their wages are paid monthly) begins with the new moon, but often before it is quite completed, they will come to their master, asking for their money, and although the month is not ended they will declare it is by an appeal to the fact that the moon *"inyanga file"* is dead. They cannot understand there being more than 28 days in a month. It is impossible to make them believe there are 31.

Small strikes frequently broke out over these issues, with Europeans consistently professing amazement at local intransigence. Strikers were seen as acting on their "unwritten and ignorant system of computing time in opposition to the statements of the Postmasters and the interpreters". The response? Recommendations of harsh punishments, often floggings and extended stretches in jail. For, obviously, power was on the side of the employers, with colonial authorities fully sharing their unwillingness to redefine calendar time to take account of local beliefs.

But even force did not bring about a quick adjustment on the part of the Africans involved, in a society where urbanization was proceeding slowly and where even urban Africans often maintained close contact with extended families in the countryside. By the late 19th century, some of the more enlightened colonists began to realize that possibly an educational effort, directed at boys, might have better effects. Thus, from 1894:

> If it could be explained to them that they are not engaged by the lunar month, it would save much difficulty....Many a score of good, hardworking boys found themselves landed in gaol in consequence of disagreements with their employers, caused in the first instance by their inability to reckon in their own time, and then the case is frequently aggravated by the employer being unable to explain matters in the native language.

Interestingly, even when English–Zulu translations were produced (by Europeans), they often mistranslated concepts into European time units.

Disputes also broke out, where longer-term labor contracts were involved, over the fact that Zulus had no fixed concept of a year, but rather simply two basic seasons – one rainy, where field work was vital, the other dry and wintry. One colonial magistrate noted, in 1868, that natives could not imagine a period of work service that went beyond one of the two six-month seasons. "It seems as if the kafir was unable to conceive of a longer unbroken term of exertion."

And all of this was in addition to even more predictable disputes over hours in the day, made still more complicated by a deep conviction that the nighttime was dominated by evil spirits and intensely unsafe. Again, the common European lament, that locals deeply believed that work could only start about an hour after sunrise and end about an hour before sunset (neither marker, however, defined by the clock; simply by natural rhythms). "It is difficult either to induce or compel [Africans] to work either before or after these periods of the day, which have received his arbitrary definition of sunrise and sunset." Problems were especially acute in the African summer, when daylight often lasted only about ten hours.

Particularly through the middle decades of the 19th century, various European employers struggled with this timing issue in addition to the others. Sometimes they simply could not find workers willing to meet their requirements of timing and duration. A few tried to resort to piece-rate agreements: instead of time-based wages, assigning a measurable body of work that had to be completed, as the basis for pay. But, as we have seen in Europe, this arrangement was vulnerable when there were hopes to increase production rates. Other employers simply resorted to importing indentured workers from India, who had less opportunity for resistance and may have brought in a different time sense to begin with. And some employers merely gave in to the inevitable, and accepted the local African definitions of time because change seemed so challenging.

Even colonial officials ventured some compromises. In 1894, in South Africa, they passed a Master and Native Servant Law whereby the work year was divided into 12 standard months of 30 days each. But this had only modest impact in a situation where time highlighted a fundamental culture clash that proved difficult to bring to a close.

Problems were particularly great in the countryside, around industries such as sugar cane production, where workers were often seasonal in any event and had regular contact with their traditional rural communities. What was striking in this deep-seated conflict was, first, the many aspects of time that were involved, and, second, how long it lasted – despite the imbalance of power in favor of the Europeans. Hence another European lament: "Our initial difficulties in regulating their hours of labour have not yet been overcome, notwithstanding half a century of experience acquired in prisons, garrisons, railways and mining camps."

This kind of conflict was not the only pattern that developed in Africa during the second half of the 19th century, even in Natal. Some encounters over time were brokered more easily, often in ways that recalled experiences in the European adjustment process. And some interactions, particularly in the cities, clearly signaled a process of change, however grudging.

Diamond miners in Portuguese-controlled Mozambique, also in southern Africa, faced a familiar industrial time framework: equipment like steam engines required considerable coordination of labor; the timing of the workday was marked by whistles and sirens. Christian missionaries in the region were pushing the importance of zealous work and "orderly lives", and bells marked the time for prayer meetings and Sunday services. Christianized Africans accordingly gained a new sense of time more rapidly than others. But for everyone work in the mines was rather highly paid, and many workers marked their prosperity by buying items like watches, even though they could not yet themselves tell time (and also pens before they knew how to write). A sense of time was spreading, and when workers visited their rural villages of origin they helped spread the word even by just showing off their consumer gains. Adjustment was also facilitated by the decision, in 1892, to move to three 8-hour shifts, rather than two of 12 (a move that occurred earlier, chronologically, than comparable gains for miners in Europe). Adjustments to imposed clock time surely varied, and the process was not easy for those involved – but there was change, amid more facilitating factors than had prevailed in nearby Natal.

Even in Natal, however, there were signs of conversion, though primarily in the cities. The spread of missionary schools provided a growing number of urban Africans with training in clock and calendar time during childhood, around a devotion, as one missionary put it, to inculcating "the value of time". Observers noted a "craze for education", even among many young women, that prompted acceptance of new time schedules. Urban governments got in on the act, in familiar ways. The city of Durban, for example, established a public clock in 1860. Various centers – both factories and churches – increasingly used bells to denote time. An Anglican cathedral bell rang "precisely at nine o-clock solar time". In 1854 the Durban mayor, bent on establishing a common local time, instituted the practice of hoisting a flag at five minutes before nine each morning, and this was widely visible. Another Natal city fired off a cannon at eight in the morning, when everybody should be at work. Various employers insisted that their workers use the sound of the cannon, rather than more personal arrangements, to get started on their jobs, and a new Zulu expression for the gun, *ubain-bail,* spread widely across Natal to designate starting time. Bugles in the military camps, sounded at regular times, provided another common indicator. Several cities also established curfew times, usually at nine in the evening, when bells also sounded.

Furthermore, particularly among Christian converts, Africans, in Natal and elsewhere, increasingly became accustomed to the week, again coining distinctive terms for each day – as with Monday, "back-to-work day". Once they adjusted, the week had certain advantages, particularly in providing a regular day – Sunday – with no assigned work and, increasingly, with Saturday afternoon off, a bit of a weekend as well.

Other changes facilitated adaptation. The spread of electricity in the cities helped dim fears of night work. The new time sense could be used in protest, for example against work assignments on Saturday afternoon, or in requests for overtime pay. In 1893 a group of African workers successfully sued for overtime rates when forced to work late on Saturday, and the colonial court agreed that this accurately reflected town custom. Urban workers also became accustomed to precise times for certain leisure activities, another important change.

Overall, African patterns during the long 19th century suggest a mixture of common trends, initially imposed from the outside, and distinctive hesitations based on very different regional timing traditions. Change was definite, but gradual, sometimes incomplete, and often accompanied both by individual resistance and considerable punitive force applied by employers and colonial authorities. Rural areas were affected very slowly if at all, and frequent migration to and from cities undoubtedly complicated the adjustment process. Timing issues did not clearly become associated with nationalism, as occurred in India, but there were certainly some similar complexities overall.

A reasonable judgment about global patterns in time during the long 19th century would probably emphasize how much had changed. So many actors had reasons for wanting change, and not just Western imperialists and greedy businessmen (though they were definitely involved), or even moralistic missionaries. Ruling groups in several places, such as Japan and the Ottoman Empire, had their own reasons for pushing new calibrations of time, though they were linked to Western pressure.

Some of the results of the interaction between "time reformers" and ordinary people mirrored experiences in the West, which is a good reminder that world history is not just about the differences among key regions. Many conditions were widely shared: the time requirements of railroads, for example, did not differ much from one place to another. Many workers, in mines and factories, resisted some of the new time systems in similar ways and for similar reasons.

But there were significant regional distinctions as well. Earlier time cultures varied, and this mattered in the encounters with pressures to change. Issues for many Africans were different from those in Japan or even India. And it mattered considerably how, and by whom, change was presented, and here the factor of outright imperialism was a clear variable. Further, differential results of encounters over time might persist well beyond the long 19th century.

We know more, of course, about the involvement of reformers and Westerners than we do about the regional populations on which they sought to impose new kinds of time. There was no embarrassment about claiming the superiority of "modern" time or castigating local traditions. But the position of many resisters was clear enough, as they sought to defend a variety of different time patterns in their confrontation with foreigners and local elites.

The global experience of time in the long 19th century, clearly pushing toward greater uniformities, raises two final questions. First: to what extent would different regional patterns lead to more durable distinctions among time cultures around the world – an issue clearly raised by the comparison between India and Japan. Second, to the extent that traditional time systems had to be sacrificed, how much did this add to the burdens of modernization in general?– a question that also figures in the modern history of the West. Time and globalization were closely intertwined, and this linkage would persist later in the 20th century and beyond.

Further Reading

The most important overall survey is Ogle, Vanessa. *The Global Transformation of Time: 1870–1950*. (Cambridge, MA: Harvard University Press, 2015).

See also Ogle, Vanessa. "Whose Time Is It? The Pluralization of Time and the Global Condition, 1870s 1940s." *The American Historical Review* 118, no. 5 (December 1, 2013): 1376–1402; and Luchien Karsten, *Globalization and Time* (London: Routledge, 2013), which covers an extensive chronology.

On calendar development:

Duncan Steel, *Marking Time: The Epic Quest to Develop a Perfect Calendar*. (New York: Wiley, 2000).

On Japan:

Uchida, Hoshimi. "The Spread of Timepieces in the Meiji Period." *Japan Review* 14 (2002): 173–92.

On the Ottoman Empire:

Uluengin, Mehmet Bengü. "Secularizing Anatolia Tick by Tick: Clock Towers in the Ottoman Empire and the Turkish Republic." *International Journal of Middle East Studies* 42, no. 1 (2010): 17–36.

Tanyeli, Ugur. "The Emergence of Modern Time-Consciousness in the Islamic World and the Problematics of Spatial Perception," Cynthia Davidson, ed., *Anytime*. (Cambridge, MA: MIT Press, 1999).

On India:

Kerr, Ian J., ed. *Railways in Modern India*. (New Delhi; New York: Oxford University Press, 2001).

Masselos, James. "Time and Nation," in Sujata Patel, Jasodhata Bagchi and Krishna Raj, eds., *Thinking Social Science in India: Essays in Honour of Alice Thorner*. (New Delhi: Sage, 2002).

Morris, Jan, and Simon Winchester. *Stones of Empire: The Buildings of the Raj*. (Oxford: Oxford University Press, 2005).

Perrett, Roy W. "History, Time, and Knowledge in Ancient India." *History and Theory* 38, no. 3 (1999): 307–21.

Saha, Arunoday. "Traditional Indian Concept of Time and Its Economic Consequences." *Project Appraisal* 5, no. 2 (June 1, 1990): 113–20.

Salamé, Richard. *Clocks and Empire: An Indian Case Study*. (Providence, RI: Brown University, 2016).

On Africa:

Adjaye, Joseph K. *Time in the Black Experience*. (Westport, CT: Praeger, 1994).

Atkins, Keletso. *The Moon Is Dead! Give Us Our Money*. (Portsmouth, NH: Heinemann, 1993).

Atkins, Keletso E. "'Kafir Time': Preindustrial Temporal Concepts and Labour Discipline in Nineteenth Century Colonial Natal." *The Journal of African History* 29, no. 2 (1988): 229–44.

Comaroff, Jean. "Missionaries and Mechanical Clocks: An Essay on Religion and History in South Africa." *The Journal of Religion* 71, no. 1 (1991): 1–17.

Harries, Patrick. *Work, Culture, and Identity: Migrant Laborers in Mozambique and South Africa, C.1860–1910*. (Portsmouth, NH: Heinemann, 1994).

7 Time in the Past Century

Two rather disparate examples from the past century:

1. Novelist Virginia Woolf, in her book *Orlando: A Biography* (1928):

> But Time, unfortunately, though it makes animals and vegetables bloom and fade with amazing punctuality, has no such simple effect upon the mind of man. An hour, once it lodges in the queer element of the human spirit, may be stretched to fifty or a hundred times its clock length; on the other hand, an hour may be accurately represented on the timepiece of the mind in one second. This extraordinary discrepancy between the time on the clock and the time in the mind is less known than it should be and deserves further investigation.

One of the newer elements in the history of time during the past century is a growing perception that clock time is not the most important human reality. Not only novelists like Woolf, but also psychologists began inquiring about why time seems so variable in actual human experience, depending for example on whether one is frightened or bored. By extension, if the human reality of time is different from clock time, perhaps the modern emphasis on clock time is distorting the quality of human life. This subversive theme, the need to work around clock time, might show up in a variety of ways – including, for some, a revival of interest in certain drugs, like marijuana, that measurably slow up the experience of time. By the early 21st century, growing interest in meditation practices – which, research showed, also slowed the perception of time, might relate to ways to address certain mental disorders. The idea of time as a subjective variable added an intriguing complexity to modern ideas about the importance of time in human schedules.

2. Beginning in the 1960s, under the impetus of widely-read leadership authors like Peter Drucker, a new interest in what was called *time management* began to gain increasing attention. First focused on the workplace, widely publicized guidance in this genre spread also to private life and education. Time management explicitly recalled earlier steps in modern history, such as Ben Franklin's homilies, but it proclaimed a broad, research-based mandate for the present.

> Time management is the process of planning and exercising conscious control of time spent on specific activities, especially to increase effectiveness, efficiency,

> and productivity. It involves a juggling act of various demands upon a person relating to work, social life, family, hobbies, personal interests and commitments with the finiteness of time. Using time effectively gives the person "choice" on spending/managing activities at their own time and expediency.
>
> It is a meta-activity with the goal to maximize the overall benefit of a set of other activities within the boundary condition of a limited amount of time, as time itself cannot be managed because it is fixed.

The history of time in the past century, from the mid-20th to the present, raises one overarching question: had the big adjustments already occurred? Many historians, like many members of the contemporary general public, are happiest when they can point to revolutionary transformations, and it is always tempting to assume that recent history is filled with fundamental change. In the case of time, however, it is really possible that the biggest shifts occurred in the early modern period (mainly for the West), and then the long 19th century for the world more generally. Of course there were loose ends to clear up, fuller conversions to register, but many directions were set. This included not only the conversions to new kinds of time awareness but also the intriguing regional variations around these trends.

Yet the past hundred years have filled with a wide range of specific developments. New intellectual constructions of time, such as the Einsteinian idea of time as relative, coexisted with even greater awareness of the long reaches of historical time. Clocks and watches changed. Though the digital revolution was not as fundamental, in daily time-keeping, as the advent of more reliable mechanical clocks and watches (now called analogs, in contrast to digital), it did represent change. New devices infused the home and automobile with more timing mechanisms than even the most ardent clock buff could have imagined. And on the global stage, while imperialist commands about time lost their sting with decolonization, there were still some interesting debates about world systems and the possibility of reform.

Here's the main point: the probability that a number of basic contours were already set by the 1920s. and often before, must be juggled against the variety of new dimensions – including some modern rather than traditional concerns about time's rigidity – that unfolded over the past several decades. This is the fundamental tension that provides some coherence to the contemporary history of time.

One way to present the problem of how contemporary patterns relate to the recent past, before we dive into details, might be: considering the major features of time today, in one's own society and around the world, what are the big differences from the patterns of the early to mid-20th century? Are we talking about tweaks or major departures? We so often assume that our lives are fundamentally different from those of previous generations, if only thanks to the march of technology: but might time, still hugely important, be an exception?

Lest the many specific adjustments seem confusing, three basic subthemes can be suggested. First, there was a real revolution in scientific conceptions of time and in aspects of time technology, for example with the advent of atomic clocks in the 1950s. These changes must be considered, but it is not clear, first, that they had major impact on the ordinary experience of time and, second, that they all in fact pointed in the same direction.

Second, without question, many of the trends of the previous two centuries not only continued, but intensified. Quick example: where in 1900 it was still necessary to consider how many people had clocks, by 2000 not only virtually every home but even every separate hotel room had one. Does acceleration of an existing trend count as major change, or can it be situated as part of an established basic pattern? It is also worth noting that moralistic sermons about time clearly declined, another probable sign that major modern patterns had been widely accepted; it was less necessary to preach about them.

Third, within the established modern context for time, a variety of time cultures continued to flourish. Furthermore, there were some interesting efforts to modify the hold of time, both in general – as a number of observers, including historians, worried about the impact of time's rigidities – and in specific daily practice. How can these variations and countertrends be assessed against the continuing, and essentially global, grasp of modern timing?

The Revolutions: Science and Technology

Relativity

The most revolutionary single development in the contemporary history of time was the concept of relativity in physics, which replaced the traditional Newtonian idea that time was an absolute with the radical notion that time varied depending on speed and the presence of nearby matter. The theory, fleshed out by Albert Einstein after 1905, built on developments in physics and mathematics over several preceding decades, including explicit argument that time is relative by the French scientist Poincaré. But it may also have been connected, charmingly, with a more historical iteration of time.

For Einstein in 1905 was working as a patent clerk in Bern, Switzerland, having failed to find a teaching job after a spotty performance at the University of Zurich. The work itself was relevant, since many of the patents he handled, it being Switzerland, involved clocks and watches, and some were designed to improve the synchronization of time on railways. But there may have been a more explicit stimulus. Daily the young scientist took a streetcar past one of the more famous medieval (early 13th century) clock towers, complete with a parading jester and dancing bears when it sounded the hour. One day the sight of the clock advanced Einstein's musing about time, as he speculated about what would happen were his streetcar to move away from the tower at the speed of light: he realized that his watch would still be ticking, on the streetcar, but if he looked back at the clock it would seem to have stopped. Time, in other words, was a relative quality, and his theory – first ventured just six weeks later, more elaborately stated a decade beyond – would prove fundamental not only to theoretical physics, but to calculations in astronomy, nuclear development, and space travel.

The idea that time is relative to the position of the observer – with time moving more slowly at faster speeds, was a dramatic departure not only from Newtonian physics, which had assumed that time was an absolute, but from the whole thrust of modern Western insistence on a single standard of time. An event might occur at one time for one observer, at another for someone else in a different rate of motion. The related belief in a space–time continuum posed a similar challenge to conventional thinking. The larger theory of relativity, of which the observations on time were only a part, as it emerged more fully in Einstein's work in succeeding decades, had many other facets. The argument about the relationship between mass and energy, captured in the famous formula $E = mc^2$, would prove fundamental to nuclear research and development, which, in turn, rightly or wrongly, made Einstein particularly famous.

In terms of time specifically, Einstein's work, quickly embellished by other researchers, had several implications. First, it dramatically affected astronomical observation. Researchers now could take into account the effects of the earth's spin in distorting space/time around it. As travel in space began to be contemplated and then brought to pass in the later 20th century, allowing for the effects of speed and distance on time proved absolutely crucial.

Also in the later 20th century, allowing for the relativity of time was essential in programming communications and positioning satellites, in their application to areas such as global positioning systems. GPS satellites, traveling more than 12,000 miles above the earth at speeds of 6,000 miles per hour, operated at different rates of time from those calculated on earth. If instruments did not take this differential into account, positioning would quickly become inaccurate; an estimate of half a mile, for example, would swell to six miles in a single day. Building relativity into measurements like mapping in turn had direct impact on daily life, as people became increasingly dependent on new devices. The same new kinds of calculations of space–time also explain why, by the 21st century, precision bombing could be directed from long distances away from the target.

The relativity of time, in other words, generated new kinds of formulas, and instruments to implement these formulas, where any factors of speed and distance were involved. This had the somewhat ironic effect of requiring greater attention to accuracy in the measurement of time than had previously been the case – with major implications for time technology. Most people might be unaware of the new complexities involved, but by the late 20th century they were routinely using equipment that had these factors built in.

The other implications of relativity applied to time were more diffuse, though they have inspired considerable speculation. Did the idea of time as relative, coupled with Einstein's unintended fame as the leading genius of the contemporary world, alter concepts of time for people more generally? Certainly the notion of time as relative influenced a variety of philosophers and other intellectuals, perhaps encouraging the new kinds of ruminations about the personal experience of time that showed up in the works of authors like Virginia Woolf. There is no question that intellectual and artistic representations of time, by the 1920s and 1930s, were different from their counterparts in the 18th and 19th century, where time was seen as a more objective phenomenon. In another major contribution, deeply anchored in time, James Joyce's *Ulysses* (1922) structured the presentation by the time markers of a single day, indicated by public clocks and bells, but explored the very different experience of time in the dreams and consciousness of its characters. Literary fascination with time was taking novel shapes. All of this added to the cultural context around the idea of relativity itself.

And there was more. In what was undoubtedly the most important 20th-century artistic representation of time, initially exhibited in 1932, the surrealist artist Salvador Dali highlighted several melting clocks, draped over buildings and trees. The famous painting, repeated several other times by the artist, deliberately symbolized the relativity of space and time, the collapse of notions of a fixed cosmic order. Dali himself, to be sure, asked about the link to Einstein, claimed that his watches had instead been inspired by visions of Camembert cheese, melting in the sun. But this was not the way the painting was interpreted: it seemed yet another statement of a new view of time, of the subjective experience of personal time in sleep or the perceptions of a dreamer.

Philosophers and psychologists became deeply engrossed in discussions of the subjective nature of time, both in reaction to the idea of relativity and based on developments in late-19th century science. William James discussed the differences between the experience of stream of consciousness in individuals, and any precise clock time. Henri Bergson used presentations of time in early movies as examples of how time experience and time measurement might differ: the notion here was that a two-hour movie exposes us to a vast array of fictional time; which is more real? Clearly, interest in the new problems of defining time was widespread in the Western intellectual world.

A few scientists, and even more science fiction writers, were also inspired to speculate about time travel, which might add to the sense that time, in various forms, was up for debate.

Some observers have gone on to argue that people more widely were inspired to raise new questions about time and about certainty in general, thanks not only to Einstein but to the other intellectuals and artists who were unquestionably changing the representation of time. The proposition is hard to prove for, as we will see in many other respects, dependence on fixed ideas of time was actually increasing – as in the time management movement. Revolutionary as the relativity theory was, it did not turn back the powerful modern trend of regulating ordinary life by the single standard of the clock. There is a gap here between the excitement of contemporary science and the humdrum but pervasive kinds of time that continue to affect growing numbers of people around the world.

The idea of time as relative transformed key branches of science; it promoted new technologies designed to measure the intricacies of time more accurately; it had some wider intellectual spillover. It should be considered, finally, in any overall assessments of the key trends in the perception and use of time in daily life.

Technology

The big development in technology, in addition to steady improvements in the capacity to build clocks and watches and sell them cheaply and widely, was the capacity to measure time in increasingly small increments. And this, intriguingly, intensified a process that had long been part of the modern history of time. From attention to defining days, mechanical clocks had increasingly added the focus on hours; then better clocks and watches had factored in minutes and, by the 20th century, seconds. To this progression the later 20th century now added interest and capacity in tenths of seconds, milliseconds and nanoseconds, with devices ready to measure and to utilize these units.

Interest in measuring time beneath the level of the second began to develop in the later 19th century, when the most precise clocks became capable of this kind of resolution. Problems in astronomy created new concern about finer definitions. It became increasingly clear that separate observers of the same phenomena in astronomy, with the best of will, timed events slightly differently. This came up, for example, in several late-century expeditions designed to chart transitions of the planet Venus. The gaps, though small, resulted from different individual reaction times and personal psychology. Efforts to use the new facilities of high-speed photography to resolve differences soon revealed that different cameras, also, produced different specifics. All of this helped feed the growing interest in time's relativity, but it also called attention to the need to use tenths of second and, if possible, even smaller units, to help chart these kinds of vital differentials.

Research focus on the smallest units of time was also spurred by growing attention to electricity and light, where tiny intervals might prove essential in instrumentation. Work on milliseconds and nanoseconds (billionths of a second) developed strongly in research and electrical engineering by the middle of the 20th century. The capacity to use nanoseconds to chart the movement of electricity became vital to technologies such as microcontrollers, and also in satellite communications. New kinds of time measurement needs, in other words, flowed both from efforts to compensate for time's relativity – using small units to define differentials – and from the kinds of technologies built on electrical transmission.

All of this, in turn, called for better capacities of time measurement. Here, the most crucial development was the emergence of atomic clocks. The idea of using atomic transitions to measure time was suggested as early as 1879, but practical research emerged only in the 1930s, by using magnetic resonance as the basis for a clock. The first atomic clock was built in the United States National Bureau of Standards laboratory, in 1949, though it was not very precise. The first effective atomic clock was set up six years later, in the United Kingdom. By 1969 the technology had advanced sufficiently to allow the unit of a second to be defined, for scientific purposes, in terms of a specific atomic frequency, providing more precise accuracy than ever before. This definition was embellished and ratified under international auspices in 1997.

From the 1960s into the 21st century, atomic clocks were increasingly refined to reduce cost and, by making smaller devices, to allow wider use. In 2015 the National Agency of Space Administration, in the United States, announced plans to deploy an atomic clock in outer space, where, free from gravitational pull, it would be more accurate than any earth-bound units.

Atomic clocks came to constitute the latest stage in the long history of attempts to maximize the precision and refinement of time measurement devices. By the early 21st century most major nations maintained bureaus designed to sustain and evaluate atomic clock timing, as the fundamental standard for accuracy. At the same time, linked networks of atomic clock centers around the world set time standards globally. Finally, again by the 21st century, the clocks allowed definition and measurement of intervals of the second even smaller than the nanosecond, with new terms such as femtosecond and attosecond. An attosecond, the most ambitious reduction thus far, is to a second as a second is to 31.71 billion years.

Atomic clocks, and the kinds of time assessments that resulted, particularly came to serve in contemporary scientific research, including astronomy, and, an even older connection, in establishing best possible navigation systems, including those used in satellites. They provide vital applications on the Internet, which depend critically on frequency and time standards, and they also serve some broadcasting stations that require very precise carrier frequency. Finally, as maintained by the official agencies, they set standards for clocks in more ordinary use. Indeed, some marketers began to sell clock radios under the label atomic clocks; they were not, but the radio signals they transmitted did originate in atomic clocks.

Development of new timing devices – adding up to a substantial redefinition of the clock – and new time measurements, both emanating from highly specialized science and scientific needs, became part of the technological context of modern life. Through international coordination, they also became a vital element in the continued globalization of time, permitting unprecedented accuracy and precision in global standards and communications. They contributed to the kinds of time-consciousness that had been widely developing in the long 19th century, even though they went beyond the active involvement of most people and most of the settings of ordinary life. They unquestionably added an important contemporary dimension to the larger history of time.

Links between growing interest and capacity in measuring small units of time and the wider range of human experience also showed up in sports, though admittedly not at the level of nanoseconds. The passion for timing relevant sports events had already been established. By the 1960s and 1970s, however, record-keeping in speed sports like track and swimming began to be kept not just in tenths of second, as had been standard since the 1890s, but in hundreths – another little step toward connecting time precision with other aspects of life, with improved stopwatches part of the bargain.

The Y2K Interlude

The increasing connection between time and intricate technology generated one fascinating near crisis as the year 2000 (Y2K) approached – a juxtaposition of the fairly old Gregorian calendar and the very new information systems and storage. As they developed during the 20th century, computer systems had usually used just the last two digits of the year's date – thus, 90 instead of 1990. This was fine during the century itself, but what would happen when all four of the year's digits changed with a new Gregorian century – i.e., the year 2000. Anxious discussion began to rise from the late 1980s onward, around real fears of a systems collapse. National governments, from the United States to Japan to Uganda, devoted massive resources to studying the problem and planning correctives. Global time coordination was fully in evidence, with many official international conferences during the later 1990s. The problem was compounded by the fact that many systems had neglected the fact that while in the Gregorian calendar most new century years do not have leap years, this is not true when they are divisible by 400 – so there was a potential problem at the end of February 2000, as well.

Tension was exacerbated by a very different set of warnings, generated by a variety of fringe Christian and survivalist groups, concerned about the religious significance of a new millennium, again by Gregorian calculations. Some people expected the year to usher in the Rapture, or at least massive disruption of earthly arrangements. Small groups stocked up on supplies, the better to survive the end of the world as we know it.

Y2K, in other words, captured the relationship of time to certain kinds of religious beliefs as well as the more modern link to technology. The result was a flurry of wider public concern.

In the end, partly thanks to preparation, very little happened. Many activities were shut down for several days in case of problems. Several Delaware lottery machines stopped working; some Japanese mobile phones deleted new messages rather than old ones; in the most tragic case, misinformation about the potential birth of Downs syndrome babies was sent to 154 pregnant women in Sheffield, England. A few similar, mostly minor, problems arose on March 1.

Overall, governments and businesses spent about 300 billion dollars to prepare for Y2K. Some of this was probably unnecessary; governments that invested little, like Italy, faced no more problems than anyone else. But, in the contemporary domains of time and technology, it can be hard to know what is essential.

Closer to Home: Other New Time Technologies

On a somewhat humbler scale, three other major technological changes in clocks and watches, developing at various points in the 20th century, added further accuracy to ordinary timepieces but also increased the availability, indeed the ubiquity, of timing devices in daily life. All three developments extended some of the most basic trends in the modern history of time, now not just in the West but in most societies around the world.

The first development was the least elaborate: the introduction of electrical clocks. Experiments with electric clocks started as early as 1814, in Great Britain; early versions were all battery-powered. Experiments continued, and the first patent was issued in 1840. By this point, the devices were using an electrically-powered pendulum, rather than weights or springs, to operate the clock face. Large-scale production began in the late 19th century. Then, early in the 20th century synchronous clocks, keeping time from the oscillations from the electric power grid, rather than any internal mechanisms, were introduced. By this point, use of electric clocks was becoming increasingly common.

The big advantages of these new devices were, first, they could be produced quite cheaply, because there were no elaborate inner workings. A related plus was the absence of much need for upkeep, including regular winding – with the huge exception of the need to adjust in situations when the power went off. Public clocks, particularly, had continued to require quite a bit of labor, often daily, which electric clocks largely eliminated. The clocks were also quite accurate, assuming that the frequency of the electric current was monitored. By the later 20th century, European countries checked this aspect of the grid daily, to make sure the number of cycles every 24 hours was correct. American utilities corrected their frequencies when the cumulative error had reached three to ten seconds, though they periodically sought to get out from under this obligation (which in turn would require more frequent manual corrections to the timepieces).

Progressively through the 20th century, relatively cheap electric clocks quietly transformed many public spaces, as well as gaining ground in private homes. Schools and colleges, most obviously, installed electric clocks in most classrooms. Banks and other commercial establishments widely adopted the clock, as the centrality of more classic public timepieces declined. While some clocks were installed in cars as early as the 1900s, the big turn to car clocks occurred in the 1930s and 1940s, again on the strength of improved electric devices that required no winding. In the post-World War boom of consumer durables, clocks as well as other timing devices began to be added to ovens and other kitchen equipment. The spread of microwave ovens (first invented right after World War II but then popularized for home use from 1967 onward), constituted another connection between electricity, on the one hand, clocks and precise timing on the other.

The main point is clear: electric clocks increased the ubiquity of timepieces in all sorts of spaces in the home and in public venues. Their capacity to designate not only time but timing, as in the microwave, correspondingly intensified this crucial aspect of modern time-consciousness.

A related change, taking shape from the 1920s onward, was the provision of time-checking services by telephone companies, in the United States and many European countries. A single call would produce up-to-date information on time and often weather, and the telephone services took great pride in their accuracy. In the United States, the recorded voice of the person who became known as the "telephone lady" actually gained some notoriety, providing opportunities for the woman, Jane Barbe, in some unrelated advertising efforts and vocalizations. This service phased out in the early 21st century (ending fully by 2007), because thanks to additional new technology other kinds of time checks became so widely available.

The advent of quartz technology, the second main shift in popular timepieces, began in fact in the quest for greater accuracy in scientific timing, prior to the advent of atomic clocks. As a mineral, quartz can be made to resonate by an electric signal, and it is also quite resistant to changes in temperature. and hence maintains accuracy. Exploration of quartz's qualities first centered on uses as radio transmitters, early in the 20th century. But a similar concern for accuracy prompted the construction of the first quartz clock, in 1927, at the Bell Telephone Laboratories. During the next three decades, quartz clocks spread widely to provide greatest possible precision in various research labs. Their bulky size, however, precluded their general use. American time standards were geared to quartz clocks between the 1930s and the 1960s, when they were replaced by atomic clocks.

The role of quartz in the wider contemporary history of time began in 1969, when the Japanese company Seiko introduced the quartz watch, after 11 years of research. Here was a huge leap in the accuracy of watch technology and not coincidentally a major boost for Japan as a source of timepieces around the world. The new watches were accurate to the level of erring by only one minute in an entire year. By the 1980s quartz technology had taken over fields such as kitchen timers, alarm clocks, and other devices, replacing earlier mechanical balance wheel movements and therefore eliminating all internal moving parts – and in the process causing shock waves across the whole clockmaking industry. Not only were they more accurate than mechanical timepieces, they were also more rugged, requiring far less frequent repair. Here was another major step in the quality of timing available to a wide public.

The final step in the evolution to date of common clock technology involved the introduction of digital devices, that display time numerically rather than by the classic clock face deployed in analog units. Here, as in so many areas, initial research goes back to the 19th century, with the first digital clock developed in Germany in the 1880s: two windows displayed the hour and the minutes, respectively, with a conventional hand to designate seconds. Digital principles were quickly applied to some pocket watches, but the first digital wrist watch did not make an appearance until the 1920s. The first patent for a digital alarm clock, invented in the United States, was issued in 1956. Obviously, this suggests a steady but rather gradual process of development.

Mass production of digital watches and other digital timepieces began in the 1970s and, though the units were initially quite expensive, a growing market emerged. The only outright liability of the devices, shared by all electric clocks, was the vulnerability to power failures (including expiration of batteries), after which they had to be reset. By the 21st century some backup battery devices or automatic reset procedures sought to address these problems. As their costs declined, digital displays became increasingly common, not only for personal timepieces and alarms, but also car clocks and kitchen devices.

But the full flowering of the digital timepiece came with the emergence of the smartphone, from the mid-1990s onward, which linked the mobile phone with the Internet and routinely displayed the time. Here, accurate time was available simply by opening the screen. By the 21st century, as its popularity soared despite considerable expense, the smartphone became the leading transmitter of time to many individuals – even if their main interest, usually, was far less in clocks than in the growing variety of apps.

The spread of digital clocks and particularly the smartphone threatened to displace more conventional sources of time. Global watch sales fell by about 40% between the 1990s and 2010. though a study in 2016 claimed that sales of watches, at least in places like the United States, were finally steadying (whether digital or analog). Adaptation among young people was particularly rapid – 52% of American teenagers had watches in the1990s, only 25% by 2010; members of the Baby Boom generation, older but still working, led the pack in watch use by this latter point. But none of this necessarily meant a change in time-consciousness, since people who relied on smartphones consulted them frequently in a given hour and could not help but notice the time. And a possible surge in smartwatches (owned by 2% of the American population by 2016) might also affect the equation.

For some young people, the digital revolution reduced or even eliminated the capacity to read a clockface, a significant problem in settings such as examination rooms, where only a wall clock was available to tell the time. Some commentators noted that digital dials, though easier to read than analogs, actually required more arithmetic to do other calculations, such as figuring out what time it would be in another hour and a quarter; compared to the modern analog watch face, it also made it harder to determine the passage of seconds. While most smartphones have a stopwatch app, simply charting the passage of seconds requires special arrangements – an interesting small retreat in a period when, on the whole, interest in precision was rising. Whether the rise of the digital significantly affected modern time sense, one way or another, is at least open to debate. What was clear was that the spread of digital devices continued the historical pattern of making timepieces easier to access, more ubiquitous and more accurate; it added to the substantial evolution of everyday time technology in the contemporary period.

Digitization and the spread of the Internet, after 1990, had one final consequence: the spread of applications under the general heading of "time-tracking". These apps, on smart phones and personal computers, allowed individuals to chart time use in various activities, sometimes related to the wider advice being issued by time management enthusiasts. But they also allowed employers to calculate and check on the time use of employees – often replacing the old time clock techniques with the added advantage of allowing supervision of timing off-site and also, potentially, with increasing precision. Some time-tracking apps proclaimed their ability to measure employee time "by the second". Here was yet another intensification of trends long associated with modern work.

Time for Everyone: The Timex Example

Closely related to the new technologies that increased the accuracy and availability of clock time around the word was the steady improvement in manufacturing and sales techniques that made clocks and watches routine items for many people, particularly but not exclusively in industrialized societies. This too continued and accelerated trends that had been well established in the 19th century, linking the contemporary experience of time to its larger modern history.

The Waterbury Clock Company had been established in Connecticut in 1854. It was one of the leading American companies, a few decades later, that introduced mass production techniques to turn out inexpensive pocket watches and clocks, using a mainly female labor force preferred for their "slender fingers" and "delicate manipulation" – and lower salaries.

During World War I, pressed by the needs of the American army to have an easily visible timepiece for artillery gunners, the company modified the classic women's wrist watch with a strong cloth strap and illuminated dials, so it could be used at night. The transition from pocket watches to the more easily visible wrist watches, for men, proceeded rapidly from that point. The company also targeted products aimed at medical personnel, including nurses.

Hit hard by the Depression, the company rebounded in the 1930s by producing the famous Mickey Mouse watch, which ultimately contributed to the growing use of watches by children. Here was another sign that timepieces were becoming an even more standard part of mass consumerism.

But the biggest step occurred in the 1950s, when the company was reconfigured as the Timex corporation. Further automation and design changes, aimed at simplifying the production of Swiss-type watches, generated higher production at lower cost, while use of new metal alloys for the casings provided great durability. Never before had watches been so cheap, and so hard to break. But the company also claimed its links to "technological expertise and innovation". More tradition-minded jewelers opposed the new watches, which greatly reduced the jeweled components of the classic watch, but sales outlets instead concentrated in drug stores and other mass-consumption sites. And the watches were widely advertised by one of the most famous newsmen in the United States and also subjected to staged torture tests in TV ads, under the slogan, "Takes a licking and keeps on ticking." By 1962 a third of all the watches in the United States were Timex, and export sales soared rapidly as well.

The Timex heyday ended in the 1970s and 1980s, when cheap watches made in Asia began to flood the market (another side of the globalization of modern time). More attention to quality and fashionable design kept the brand alive, however, and there were also new opportunities to produce sports watches for triathletes and other participants in rugged outdoor competitions. By the 21st century watches could also be combined with GPS services and with heartrate monitors for use in exercise. At the same time, most actual manufacturing shifted to other, lower-wage locations, with a major center for example in the Philippines, though the company began manufacturing one product line again in the United States in 2019 – the first time watches had been made in-country in several decades.

Timex constitutes only one of many important production stories in the contemporary history of time, but it vividly illustrates the ways manufacturing and imaginative specializations made time more widely available, and more connected to a variety of human activities, than ever before.

Policy and Globalization

A number of interesting policy decisions about time emerged at various points during the past century, and in many different places. They were not as fundamental as the policy moves of the 19th century, and they often built upon them, particularly as shared standards about time began to spread even more widely on a global basis. But there were some intriguing alleyways. In some ways, as well, potential reform options were turned away, as the keepers of modern time perhaps became a bit conservative in some respects.

Calendars

The big news was fairly predictable, given prior trends and the steady advance of global contacts and the attendant need for better coordination. Use of the Gregorian calendar spread steadily. Eastern Europe made the move fairly early. The Russian revolutionary government adopted the calendar in 1917, which meant that their October revolution was now attached to early November. Romania, Bulgaria, and then Greece all accepted the Gregorian calendar by the early 1920s, completing its currency in Europe – though in no case did the Orthodox Church follow suit, retaining the Julian system. Chinese revolutionaries made their move in 1912 – again demonstrating how alignment with Western/global standards still seemed to stand for progress. In fact, civil war ultimately further locked China into the conversion to the Gregorian system. Nationalists had sought to combine the new calendar with an older monthly numbering system, but the communists in 1949 simply adopted Western-style numbered years. Of course the older lunar method still defined many traditional holidays. Several southeast Asian countries began to combine the Gregorian calendar with a second, more traditional system, such as a Burmese calendar in Myanmar. Israel would do the same, using both Gregorian and Hebrew standards. Most Middle Eastern countries came to use a Gregorian calendar for business and political purposes – Saudi Arabia made the move in 1961, while maintaining the lunar system for the Islamic religion. Only Iran, Ethiopia, and Afghanistan continue to hold out entirely.

India went through a particularly fascinating debate in developing a separate national calendar alongside the Gregorian – a revealing combination given the earlier tensions with colonial impositions. Shortly after independence, the new government established a Calendar Reform Committee, in 1952. The goal was to develop an appropriate national system to be used alongside the Gregorian, and to touch base with older Hindu calendar traditions. The committee noted that there were 32 local systems in use at the time. Eminent scholars chaired the effort, which was meant to establish a scientifically accurate system that could be deployed throughout the nation. Prime Minister Nehru noted that the different existing calendars

> represent past political divisions of the country….Now that we have attained independence, it is obviously desirable that there should be a certain uniformity in the calendar for our civic, social and other purposes, and this should be done on a scientific approach to this problem.

The result is a year that begins in late March (in Gregorian terms), with 12 months, five with 31 days, the rest with 30. A leap year applies to the first month (Chaitra), every four years. Names of the months derive from older Hindu terms. The Saka era is used instead of BC/AD, and begins in year 78 of the Gregorian calendar. Again, this national calendar is used alongside Gregorian standards, keeping India in full contact with what had become the basic global system but with the opportunity, as well, to claim a separate calendrical identity.

Along with the various combinations and adjustments, the big calendar news was the continued refusal to entertain, on any global basis, more dramatic reform proposals. Since the 19th century a variety of schemes have been devised to repair the many flaws of the Gregorian system: the fact that the months vary so much; that each year starts on a different day, which in turn muddies the scheduling of key holidays and makes it impossible easily to predict the weekday of any date in the year; that the quarters of the year are uneven in length, complicating accounting systems. Ideas for change won a considerable audience between the world wars, and efforts resumed after World War II, backed by several ad hoc organizations. By the 1950s relevant United Nations leaders came to prefer one approach, called the World Calendar. With this, each year would start on a Sunday. There would follow 52 weeks of seven days each; every quarter (13 weeks) would begin on the same day. Two intercalary days at the end of the year, to bring the total to the solar pattern, would be holidays, called World Day and Leap Day respectively. But for all the reform enthusiasm and apparent rationality, nothing really has come of the World Calendar idea, or any of the several similar proposals. Many religious leaders dislike the notion of two new holidays that have no religious function. Most other world leaders, rightly or wrongly, fail to see any particular priority in this kind of change. As a result, most of the world now runs on a calendar devised in the 16th century.

Religious calendar issues also generated some discussion. A few Christians periodically wondered if the date of Easter could be regularized. Some Islamic leaders debated the possibility of determining the timing of Ramadan in advance, by scientific calculation, rather than depending on local determinations or observational results in Saudi Arabia – for there was a recurrent history of some dispute over this issue.

Finally, on the fringes of policy but closely wrapped up in globalization, a growing number of societies outside the Christian tradition, but converted to the Gregorian calendar, also began to accept a dating system that posited the major divide between BC and AD. Efforts to base historical dates on a different system, derived from regional history, faded, though this remained crucial to the Indian calendar and also operated in North Korea.

Conversion here was aided both by simple habituation to the idea of common global standards and by growing use, by the late 20th century, of the BCE–CE terminology (Before the Common Era/Common Era), to replace the explicitly Christian abbreviations. These designations had been developed by European Jews from the late 18th century onward, as they increasingly accepted the Gregorian system but shunned purely Christian symbolism. Now, with growing interest in world history in many places, BCE/CE seemed a more appropriate usage.

And these changes, finally, help explain why so many parts of the world, and not just those of Christian heritage, gleefully celebrated the advent of the new millennium, as the year 2000 was rung in. (Technically, the new century and millennium did not really begin until January 1, 2001, but this was widely ignored in favor of the more obvious symbolism of 2000.) India and most Muslim countries held back, though there were festivities in Egypt; again, globalization was not yet uniformity. But the Chinese featured fireworks and dragon dances, while the communist party secretary lit an "eternal flame" and pledged that "China would recover its lost glory." Japan sponsored a series of concerts and fireworks and, at midnight, temple bells rang out 108 times across the country, "to dispel the evils of mankind". The overall result, along with the more predictable celebrations in Australia, Europe, and the Americas, all abundantly televised around the world, was arguably the greatest global party ever, entirely based on a shared sense of calendar time and a desire for spectacle.

Hours

Policy issues relating to hours of the day were even less systematic than those involving the calendar, again mainly because the big decisions, on time zones, had already been taken. There were, however, some interesting specific debates, a few of which are ongoing.

The most important remaining time zone discussion took shape in China during the 20th century. In the late 19th century, European enclaves had used a single clock standard to determine coastal time, but most of the nation, largely oblivious to the international discussion, remained locked in local zones, though in principle the solar schedule in Beijing was supposed to provide official guidance. The new revolutionary government sought to work more explicitly on the local zone problem after 1911, while also accepting the international time zone approach. Ultimately, five zones were proposed for the huge country. But for several decades the government was too weak to impose the new pattern, and local vagaries persisted. However, in 1949–50 the victorious communist regime decided on a single national zone, based again on Beijing time. As in India in the late 19th century, it was deemed more important to national unity to have a single zone than to worry about the fact that daily solar patterns would vary by more than four hours from one end of the country to another. Here, as in India, was an important contemporary illustration of the role of political power in key decisions about time – in this case, largely ignoring concerns from outlying regions. Only one province, Xinjiang, where there are considerable tensions between the local Uighur and ethnic Chinese populations, has insisted on maintaining two systems –with the preferred Uigher pattern, more compatible with the time standards of neighboring Muslim countries, two hours different from the Chinese standard. The result, familiar from other cases of dual systems, was a need to ask individuals which time they were using – a manageable adjustment, but often confusing.

Russia, with 11 time zones, periodically considered some consolidation and briefly in the 20th century cut back to nine, but later reversed this decision.

Other than residual time zone debates, the big issue, in many parts of the world, revolved around whether to adopt, retain, or reject the idea of a seasonal daylight savings time. The concept of altering the clock to reflect seasonal change –in the northern hemisphere, pushing the hour forward in spring and summer, to take advantage of extended daylight in the evening – was often presented as a dramatically new idea. In fact, it sought in a way to apply the old notion, of shifting the duration of hours by season, to the new agreement that hours should be a standard length throughout the day.

Proposals to adjust the clocks twice a year began to surface in the 19th century, though Benjamin Franklin, as ambassador to Paris at the end of the 18th century, had already satirically suggested that the French could save a lot of candles if they got up and also went to bed earlier in summertime. Increasing use of electricity, and attendant expense, provided a new spur. As early as 1810 the Spanish National Assembly decreed that certain meeting times should be adjusted by an hour during summer, but it did not propose changing the clocks. A New Zealand scientist, George Hudson, first floated the idea of a seasonal clock change, in 1895, and his idea sparked wide interest. Independently an English sportsman made the same suggestion in 1905, mainly because he resented having to cut short his golf game at dusk in summer – a different clock would give him an extra hour of daylight. Several cities in Ontario were the first actually to adopt daylight savings time (DST), between 1908 and 1912.

But the real push came during World War I, when a number of countries were desperate to save energy. Germany and Austria-Hungary led the way, but Britain, the United States, and others soon followed. Some countries quickly dropped the project after the war ended, but some persisted, including Britain. Then World War

II provided another boost, and in the decades after the war a number of regions worked to generalize the system: the United States finally adopted national legislation to regularize the clock changes (with a few states or localities dissenting), and the European Union did the same. Varying specifics complicated the system: different governments started and ended DST on different dates.

From the outset, DST prompted highly emotional debates, highlighting the fact that, even in the contemporary world, time was not just an objective item. Farmers almost always opposed the system, arguing that it disrupted agricultural work patterns and also jeopardized the safety of rural school children, forcing them to walk or wait for buses in the dark during the spring months. Theater owners often objected, worrying about attendance during extended periods of sunlight. But sportsmen liked the system (including President Woodrow Wilson, another avid golfer, who twice vetoed new laws repealing DST after World War I, only to be overridden). At one point Idaho potato growers defended DST, on grounds that more French fries were consumed in the long summer evenings. Some observers wondered why people could not simply get going earlier in the morning during summer, rather than shifting the clock. Periodically, on the other hand, suggestions surfaced that DST be retained year round, but they were often countered by concerns for children's safety and objections from more northerly regions over prolonged darkness during the winter morning hours. Russia tried this approach in 2011 but had to back down three years later, given the winter darkness complaints. Most Muslim countries stayed away from DST because of its impact on Ramadan fasting (when Ramadan occurs in the summer, DST delays the post-sunset dinner), though Iran sticks to DST even during summer fasting.

On the whole, by the early 21st century, debate continued with a slight trend now turning against the system. A rural region of Australia rejected DST after prolonged controversy around 2010. Russia, when it abandoned its DST experiment, simply shifted back to "standard" time year-round. Current arguments, in places like the United States, mirror earlier alignments: farmers still opposed, recreational outlets in favor. But there are some new points, some of them particularly revealing about human interactions with time.

Thus, the economic benefits of DST are less clear than imagined. Electricity is saved, but more heating fuel is expended when people get up earlier in spring. Crime and accidents may go down thanks to DST, but accidents at work rise during the weeks when people are first adjusting to the time changes. But the big new claims involve, first, the sheer nuisance and expense of shifting timepieces twice a year, in societies that, overall, depend heavily on time consistency. And, second, there is a growing realization that people's body rhythms do not adjust easily to clock shifts: sleep (including children's sleep) is often disrupted for weeks, and there is actually an increase in heart attacks for a few days after the spring time change (but in fall, when clocks "fall back" to standard time, heart attacks go down by almost as much, presumably because people get more sleep). The Russian government cited the "stress of changing clocks" in its 2014 decision for a single system, while Kazakhstan emphasized health concerns as its reasons for dropping DST in 2005. Growing reliance on computer-based systems also raises risks of disruption, though programs can be adjusted to make the two annual changes automatically.

Some countries, finally have cut through the problem in another way, mandating DST year round – in this camp are places like Canada, Turkey, Argentina, Morocco, and Namibia. A growing number of American states have also sought to move to this pattern, though the necessary federal legislation has not emerged. And in a revealing concession, the European Union is abandoning the twice-yearly clock change in 2021: member states can opt either for standard or daylight savings time all year. Debate continues, but the balance may be shifting thanks to a human and systems preference for regularity.

A final debate about time and the human body began to emerge in the United States and elsewhere in the early years of the 21st century, again involving no small amount of passion. Here the issue was early school start times in their relationship to the peculiar rhythms of the adolescent. From the 1990s onward, a growing group of experts, particularly in the United States and Brazil, began calling attention to the problem of sleep deprivation among teenagers caused, they argued, by poorly chosen school schedules – though it is fair to note that the problem may have become more acute with the growing distraction of smartphones and games delaying effective bedtimes. The American Academy of Pediatrics came out with a recommendation that high schools and middle schools delay start times until 8:30 or later. And a number of school systems did adjust their schedules by a half hour, though rarely seeking to meet this standard. Against the pleas for change were concerns about time for activities and jobs in the afternoon; bus schedule complications; and the well-known bureaucratic reluctance to change anything too far too fast. Here too debate continues.

On the whole, policy issues concerning clock time now worked around the edges of the modern and increasingly global patterns, not at the core. Wide though not uniform agreements had been reached on issues like time zones and calendars. But new technological systems and new data about the complexities of the human interaction with time prompted some interesting discussions and adjustments. Time, as a policy concern, does not stand still.

Time and Society

The increasing availability of timepieces, both older types and newer versions, undoubtedly reflected the fact that, for growing numbers of people around the world, clock-based schedules were becoming even more important than in the recent past. Here too, fundamental trends were not new, but there were significant intensifications; and some people, additionally, were exposed to smaller chunks of time than ever before. Several illustrations flesh out these basic points.

One caveat in advance, however: this section addresses common trends, which affected many groups of people in many regions. It does not imply uniformity, however; there will be more on variety as a complicating factor in the next section.

Children

One of the basic features of global society in the 20th century was the steady expansion of education and, with this, in most countries, a related expansion of the exposure of children to scheduled clock time. A school in India notes that a bell rings at 8:15, followed by a brief prayer and assembly, and then a final bell at 8:30 when all the students should be at their desks – reminiscent, if more benignly, of earlier factory systems. Japanese schools begin at 8:30 sharp (an interesting contrast to start times in the United States; there has been no need for debate over this particular issue in Japan), and punctuality is vigorously instilled.

For older young people, in high school and particularly college, time management instruction began to add to the inculcation of time sense by the early 21st century. A variety of college programs, either optional or required, explicitly preached time management in the United States, in the belief that many young people, particularly when freed from parental supervision, needed an extra boost in using the clock wisely.

Children in many societies were also increasingly subject to family timing. Russian parents insisted that children come home at agreed times. Many families imposed bedtimes and, for older children, curfews. Many children also adapted daily schedules to the timing of favorite radio or TV shows, though with recording devices this eased by the 21st century. Pauses that were, interestingly enough, called "time-outs" became a favored method of discipline in several countries.

Considerable attention was devoted to teaching children to tell time. Parents in the United States were told that children could begin to address time by age 3, at least in terms of some initial orientation, and could advance by age 5. During the 20th century a variety of toys were offered to aid the process, most notably model clocks where kids could manipulate the hands. In 1968 one American toy company went a step further, selling a talking clock that could instruct viewers as the clock face moved.

Schools joined in, often including clock exercises as part of basic arithmetic (and, indeed, simply teaching children to count to 60 was a vital building block, which many parents worked on even before school age). In many American schools, first graders were taught to deal with hours and half hours; second graders learned to identify five minutes and single minutes, along with the ability to calculate elapsed time forward and backward – that is, what time was it a half hour ago, or will it be a half hour later. A number of catchy dittos plus use of model clocks were meant to make children increasingly comfortable with the process. Third graders, learning to multiply by five, could understand that 5:40 meant 8 x 5 minutes after the hour. And so on.

The introduction of more widespread digital timepieces in the 21st century meant that many children stopped learning to tell clock time – though the capacity was still widely taught. Schools in the United Kingdom often increasingly conceded to digital time, more often than their counterparts in the United States. But the shift – which did leave many older children incapable of telling time in the conventional sense – did not necessarily affect basic time sense. And for many middle-class children, whose parents increasingly scheduled full days of athletic practices, music training, and so on, time scheduling if anything became more rigid than ever before. The same applied to many Asian societies – Japan, Korea, China, India – where many children attended tutorials after school time.

Working Mothers

Both national and global surveys in the 21st century revealed no gender difference in time-consciousness, which may represent a bit of a change since the long 19th century. Whatever the historical patterns, there was no question that, by the later 20th century, a growing group of women experienced a new and substantial contact with the clock.

Throughout Western society, increasing numbers of mothers joined the formal labor force from the late 1950s onward. To be sure, many women had always worked but, most commonly after marriage, either in the home or in occupations like domestic service that could be hugely burdensome but did not impose a modern time frame. So the movement outside the home was a substantial innovation. By 2019 in the United States, 70% of all mothers with children under 18 were working, 75% of them fulltime; and while percentages varied somewhat in other countries, broadly similar patterns developed.

The result was a tremendous time pressure, as working women continued to shoulder a disproportionate share of parental and household duties along with their formal work schedule. One estimate suggested they worked, at their two areas of responsibility, at least 98 hours a week; and as one American mother put it, "working fulltime while having kids is just plain hard".

In this setting a new female advice industry sprang up, offering what sounded remarkably like 19th-century time advice but adjusted to the vocabulary of contemporary time management. Make sure you use time wisely. Put a planner into effect, so you get things done systematically. Stay focused and don't procrastinate: don't put off the tough jobs, but "think of how awesome you'll feel if you complete that big task". Shift gears according to the clock. When the job is over at 5, focus entirely on family (plus a bit of time for yourself); correspondingly, on the job forget about the kids. If you can't quite do this take five minutes out to figure out what you will do when you get home, but then return to the work schedule. It's all possible, the new advice literature urged, if you just concentrate. "Finding the right way to manage your time and energy comes with practice", but happily you can master the time pressures: "This is how you'll get more done."

In practice many working mothers clearly reflected time pressures, noting that, unlike their male colleagues, they could not linger at work, rushing away from the office at 5 o'clock sharp to get their kids to whatever practices or activities seemed vital. One story suggested that the person most likely to run a red light in traffic was now the working "soccer mom", struggling to get the kids where they needed to be after her workday was done.

Recreation Time

We have seen that one of the most interesting developments in the long 19th century was the imposition of clock time on key leisure activities. This connection expanded at least through most of the 20th century. Along with theater schedules and timed sporting events, new media technologies brought new schedules in their wake.

Increasing amounts of personal and family time were now bent around the schedules of radio and then television shows.

Radio came first, of course. Regular radio shows began to emerge from 1920 onward and, as a matter of coordinating necessity, were normally rigorously timed. Daytime soap operas took off from about 1930. Prime time newscasts date from that decade as well. The popularity of key shows could attract huge audiences, eager to make sure they were ready to listen on time. Thus the "Lone Ranger" radio show, launched in 1933 and soon broadcast from 250 stations, ultimately drew an audience of millions, both children and adults, who simply built in the 7:30 start time into their daily routines. Similar time scheduling drew many housewives to the soaps during the day, as with the long-running British show "The Archers", launched in 1950 and still going strong. The soap opera genre initially featured 15-minute segments, then expanded to 30. And, while different countries featured different soap themes (interesting comparisons offer contrasts among Turkey – the largest exporter of soaps to Asia, the United States, Australia, and Russia), they all had a similar capacity to win deep personal attachments around which a housewife's afternoon might be organized. Nightly newscasts offered yet another opportunity to organize time, during or right after the family meal.

Television gained similar drawing power. By the early 1980s 27–29 million Americans were regularly tuning into the leading newscaster, Walter Cronkite, for his scheduled evening broadcast. Television soaps ultimately lost some ground, partly because more women were working outside the home, but they still drew some faithful viewers Top "prime time" shows easily had the capacity to organize part of a normal evening for many. The final episode of MASH (the Korean-war based comedy) in 1983 drew 106 million American viewers, who had carefully timed that part of their evening. At least into the 1990s, when new technologies allowed recording and greater variety diluted audiences, radio and television cemented the connection between time and recreation during much of the 20th century.

Short Time

Many people in the later 20th century began to respond to smaller time intervals than had been widely emphasized in the previous period, quite apart from the acceleration of more detailed timing for some athletic performances. That is, more occasions arose where calculations in minutes, rather than hours or half hours, gained growing prominence, building on the increasing familiarity with clock time.

Obviously, as commercial television loomed larger in many lives, understanding the duration of advertisements (as opportunities to move away from the set for a minute or two), became a part of recreational life. And TV in many countries also exposed viewers to countless game shows, where contestants were put through questions where an answer should be provided within 60 seconds or so, often with a loud clicking mechanism to heighten the tension.

Other kinds of scheduling began to be based on shorter intervals. Many schools and colleges scheduled classes at times like 9:10 or 2:40, to facilitate classroom utilization. A number of services began to use quarter hours for appointments: doctors, restaurants, car repair shops. The modern phenomenon of parking meters added yet another emphasis on the importance of minutes: the first parking meter was set up, in Oklahoma City, in 1935, and the phenomenon spread widely. A variety of exercise recommendations for busy people featured activities that could be timed for five or ten minutes inserted into the daily routine. In the 21st century, as well-being experts recommended periods of meditation (when, presumably, time sense might be relaxed), these too could be explicitly timed in doses of five or ten minutes.

A few other gestures were revealing. An organization called TED (Technology, Entertainment, and Design) arose in the 1980s, sponsoring talks on a wide range of subjects. By 2019 the network had spread to over 100 different languages, with a worldwide audience. And TED talks were rigorously timed: 10 minutes, no more no less. After 2015 another trend emerged, though on a much smaller scale: a series of competitions in various American universities called "the three-minute PhD", in which efficient graduate students presented their whole doctoral corpus in a rigidly timed 180 seconds. Not very important in itself, and often an interesting challenge, the exercise was another indication of the fascination with short intervals.

None of this was necessarily revolutionary for people already imbued with clock time, but it did advance the specifics of timing beyond what had been ventured in the long 19th century. It added to the many other extensions of clock-based scheduling during recent decades.

Variety and Resistance

The experience of time in the contemporary period of world history did not generate the kind of overt protests that had marked the later 19th century. There were no outright rebellions or attacks on clocks as had occurred in India or Africa. For better or worse, clock time had gained wider acceptance.

But responses to time still varied widely, reflecting earlier traditions and new opportunities. Patterns here require some careful assessment, lest older prejudices about time and morality resurface. But the variety was widely attested, and it adds an important, and realistic, complexity to what had already been a complex history. Three patterns stand out: within cultures, across cultures, and amid recent change.

Individuals

It is hardly a surprise that, within a given society, individuals vary widely in their approaches to punctuality, even when they share a basic time sense. What is new is the availability of more precise data, thanks to psychologists' growing interest in the relationship between time and personal psychology.

Thus a 2015 study compares watch-wearers and non-wearers in the United States – a 40–60 split in the population at large given the decline of watches in favor of other devices. Predictably, watch-wearers are far more conscientious and more likely to show up early for meetings and appointments than their bare-wrist counterparts. They are also more introverted than their watch-less colleagues, less agreeable, but more emotionally stable. Time sense, in other words, probably relates to some broader psychological variables. But the results remind us that, even in a time-obsessed society, there are many different individual patterns.

Cultures

Here is the trickiest category, again because of potential prejudice, but a vital reminder that older cultural patterns still count in a globalized world. A 1999 study thus set out to measure differences in pace of life in 31 different societies, with time as a vital component of pace. Criteria were: speed of walking; speed of completing tasks in post offices; and synchronization or lack of synchronization in bank clocks in capital cities. The results were quite clear, though perhaps a bit surprising. Nine northern European countries plus Japan headed the list for fast pace. Switzerland topped all others in coordination of time in the bank clocks, reflecting its strong role in the history of timepieces. Japanese time culture stood out, not only in this survey but also in the many accounts of the importance of exactitude in train and subway schedules, with officials offering public apologies for missing by a few seconds whether too early or too late (thus in 2017, "we sincerely apologize for the inconvenience caused", when a crack train left 20 seconds before its scheduled departure). Reports from Germany, another fast-paced center, suggested a preference for getting to meetings ten minutes early.

The United States was on the high end in the pace survey, but toward the middle of the pack (16th overall, but only 20th in clock accuracy). But attention focused strongly on the "low" group – mainly in warmer climates, less industrial, and arguably drawn into global standards both somewhat later and more reluctantly: thus, South Asia, Africa, the Middle East, Latin America. Costa Rica, alone of the Latin American countries in the group, joined the United States in the upper-middle pace ranks.

Anecdotal data tend to confirm these regional distinctions in time sense. Thus being late for an appointment in Malaysia is not only common, but does not necessarily require an apology. Brazilians stereotypically are also unpunctual, unless a meeting schedule is set according to "English time". Russians are less punctual than many northerners, though they often expect that foreigners will be on time. (But the most famous tardy Russian of the early 21st century, President Vladimir Putin, was an outlier among his own countrymen. Often very late for meetings with popes or prime ministers, possibly as a power tactic – nothing like showing that you can defy time in a time-conscious world, Putin was also notoriously late in personal encounters; though he did have the capacity for showing up on time for events like scheduled television appearances.) China operates on the high side where time precision is concerned, accepting ten minutes of tardiness but no more; but obviously less compulsive than Japan. African time remains more loosely structured.

But with all this significant variety, there are a few counters. First, it is vital to remember how individuals differ, no matter what the culture. India, for example, features one of the most famously time-conscious leaders of the 20th century. Mohandas Gandhi carried a cheap but accurate pocket watch and regulated not only meetings but personal scheduling like exercise by its count (he also gave watches as presents, even to two British security agents assigned to keep tabs on him). Believing in obligations to the nation, he urged "You must not waste of minute of your time", essentially a nationalist variant on Ben Franklin's advice.

Second, many people in modern societies have developed the capacity to adapt their time sense to circumstance, regardless of overall culture. Even in less time-obsessed countries, most people show up on time for catching trains and planes – which in turn are often on time; the subway in New Delhi, India, for example, has a 99% success rate in arrivals and departures within a minute of schedule, actually doing better than its counterpart in rush-hour Tokyo. A somewhat global time sense can reach even deeper. A recent study of several Amazonian tribes revealed that their local languages still have no word for time; rainy and dry seasons represent their principal time measurement. But they also know Portuguese and can use it to reflect clock time and 12-month calendar time; they even maintain Gregorian calendars with local artistic themes. In other words, they have two kinds of time sense and choose depending on circumstance.

Finally, comparisons may sometimes seem more straightforward than they actually are. Northern Europeans seem to have a particularly acute time sense compared to Americans, which is interesting; but they also show more capacity to adjust this sense to circumstance. Thus they are much better than Americans and Japanese at turning off work-time according to the clock, shifting into recreational mode much more fully after hours or on weekends. Thus an American accustomed to staying after normal quitting time to demonstrate his loyalty to the firm, but now assigned to a British office, was put to shame by colleagues who left the office at 5 o'clock sharp; to justify his compulsive work time habits he took to claiming that he was trying to avoid the traffic. Cultural variations in time habits are real, but sometimes surprisingly complex.

Relaxations or Resistance?

Here is the trickiest category of all, but potentially quite important. Not every major trend in contemporary world history advances the central patterns of modern timing. And some key new developments may create some important counter-pressures.

Thus the popularity of automobiles, in so many societies, has something to do with their ability to allow some flexibility in timing, compared to trains – though for many weary commuters who must hit the road early the difference may be negligible. The contemporary capacity to record media shows, as noted, significantly reduced the need to stick to a particular time schedule during afternoons or evenings, though sometimes "spoiler alerts" were essential to avoid premature knowledge. Every indication suggests that, partly due to more mothers working, the family meal time became less rigorously enforced by the later 20th century, for better or worse. Teenagers, particularly, often carved out their own patterns.

Vacations and weekends could be key testing points. The rise of computers and smartphones made intrusions into these domains increasingly easy. Many vacationers now might agree to accept a call or even a videoconference at a particular time, thus intruding into what was in principle time-free space. Many Europeans, and even some governments as in France, moved to try to protect nonwork hours from this kind of time injection. Americans and Japanese, with far less vacation time to begin with, were probably more vulnerable.

Perhaps the clearest recent partial rebellion against time centered on the hippy movement of the 1960s and slightly beyond, though the modern time sense was not one of their more explicit targets. Hippy communes notoriously ignored rigorous schedules as part of their rejection of consumer society. Hippies often hitchhiked or otherwise traveled irregularly, again avoiding precise time. Use of hallucinogenic drugs, another common hippy interest, might considerably modify any time sense. And, while popular performers could participate in carefully scheduled concerts, many hippy events – like the great 1969 festival in Woodstock, New York – stretched over several days, with people showing up as they wished and with no overarching schedule involved. Casual timing also marked hippy events in other countries, including a planned festival at Stonehenge in England that was ultimately blocked by the authorities. By the 1970s hippy culture began to fade, and its impact even on the time sense of a particular generation was not clear; but the episode is worth noting.

During the past half century, religious figures have sometimes added their voices to the criticism of the unrelenting qualities of modern clock time. Thus a Buddhist writer, early in the 21st century, urges people to take off their watches in daily interactions, as a means of resisting the pressures of small increments of time. "Buddhist time", he claims, is the antithesis of artificial clock time, allowing focus simply on the "now". Through Buddhism, the recommendation continues, people can "enter a timeless zone, nowhere to go, nothing to do, free of scheduling and appointments", in what is called "nowness awareness".

Finally, on a much different and potentially wider scale, the spread of retirement in many ageing societies raised yet another possibility of modifying the modern time sense – in this case by people who had for the most part lived by the clock but could now often put it aside. Retirees became known for what was called "time affluence", with up to eight hours a day available for whatever personal agenda made sense, one day to the next. Many, accordingly, began to limit watch-wearing, making a distinction between home and now only occasional forays into the outside world.

By the early 21st century, in part because of growing understanding of the pressures and downsides of a modern time sense, in part because of earlier questioning by intellectuals like Virginia Woolf, some people became increasingly aware of the desirability, at least periodically, of loosening the hold of time. A number of experts on well-being or creativity explicitly urged individuals to carve out periods in which time could be ignored. Along with larger developments such as ageing and retirement, the result may suggest at least some rebalancing against the predominant modern trends. Here, at least, is a final factor to consider amid the various time-related developments of recent decades.

The past century has seen a number of trends that confirmed or extended modern uses of time, in many parts of the world – largely building on previous developments but intensifying their impact. Variety has persisted as well, illustrating the balance between the regional and the global in contemporary world history. Finally, a few developments suggest interesting modifications of modern time pressures and even new challenges to any rigid definition of time itself. The result is an intriguing mix, that offers a certain range of choice; but, on balance, it highlights the steady amplification in the global role of clock-based schedules.

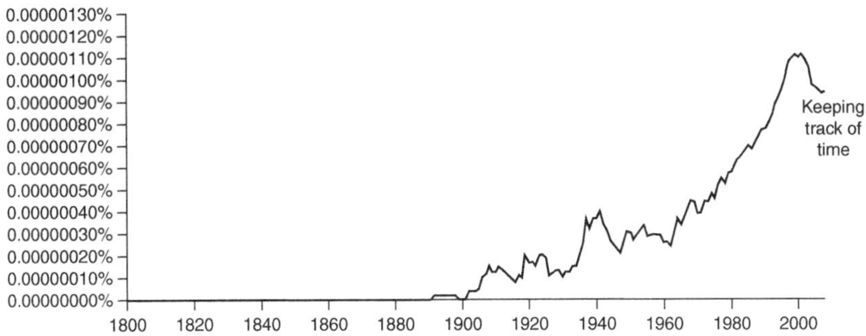

Figure 7.1 Frequency of the phrase "keeping track of time" in English, 1800–present, Google Ngram viewer, accessed August 15, 2019.

Thus, amid undeniable complexity, a final point. The relative English-language frequency of reference to the phrase "keeping track of time" shows an initial upswing in the 1940s and then a huge surge from the 1960s onward, suggesting, as do so many recent developments, a rising consciousness of time (see Figure 7.1). Google Ngram data are of course only suggestive, but this pattern certainly meshes with the other contemporary changes that can be traced in greater detail. Time awareness goes up in the later 20th century, including the sometimes rueful reminder that someone should be keeping tabs on the clock. And even when individuals "lose track of the time", another modern linguistic possibility, often offered in apology for unpunctual arrivals, they almost certainly imply that they know there is a time to keep track of.

Further Reading

On relativity:

Galison, Peter. *Einstein's Clocks and Poincaré's Maps: Empires of Time*. (New York: W. W. Norton & Company, 2004) deals with the scientific background and Einstein's work itself.

On fractions of the second:

Canales, Jimena. *A Tenth of a Second: A History*. (Chicago, IL: University of Chicago Press, 2011).

On production and consumerism:

Dwyer, Douglas. "How Quartz Watches Work." HowStuffWorks, April 1, 2000. https://electronics.howstuffworks.com/gadgets/clocks-watches/quartz-watch.htm.

McDermott, Kathleen. *Timex: A Company and Its Community 1854–1998*. (Timex Corporation, 1998).

Viredaz, Michel. “An Introduction to Electric Clocks.” Translated by David Read. *NAWCC Bulletin* (April 2003): 1–10.
Wise, S J. *Electric Clocks*. (Manchester: Heywood and Company Ltd, 1948).

On daylight savings and time reform proposals:

Bartky, Ian R. *One Time Fits All: The Campaigns for Global Uniformity*. (Stanford, CA: Stanford University Press, 2007).
Downing, Michael. *Spring Forward: The Annual Madness of Daylight Saving Time*. (Washington, DC: Counterpoint, 2005).
Prerau, David. *Seize the Daylight: The Curious and Contentious Story of Daylight Saving Time*. (New York: Basic Books, 2009).
Rakoff, Todd D. *A Time for Every Purpose: Law and the Balance of Life*. (Cambridge, MA: Harvard University Press, 2002).

On working women and time:

Agno, John, and Barbara McEwen. *Women and Time*. (Salt Lake City, UT: Signature, Inc., 2011).
Evetts, Julia. *Women and Career: Themes and Issues In Advanced Industrial Societies*. (London; New York: Routledge, 1994).

On mass media:

Buxton. Frank and Bill Owen. *Radio's Golden Age: The Programs and the Personalities*. (Norwalk, CT: Easton Press, 1966).
Rubin, Alan M., and Elizabeth M. Perse. “Audience Activity and Soap Opera Involvement: A Uses and Effects Investigation.” *Human Communication Research* 14, no. 2 (1987): 246–68.

On variety:

Ellis, David A., and Rob Jenkins. “Watch-Wearing as a Marker of Conscientiousness.” *PeerJ* 3 (August 27, 2015).
Sinha, Vera da Silva. “Event-Based Time in Three Indigenous Amazonian and Xinguan Cultures and Languages.” *Frontiers in Psychology* (2019).
Zimbardo, Philip G., and John N. Boyd. “Putting Time in Perspective: A Valid, Reliable Individual Differences Metric.” *Journal of Personality and Social Psychology* 77, no. 6 (1999): 1271–88.

8 Conclusion

There is a substantial gap, and not just chronologically, between the "first clock" in Scotland or notched sticks in the Congo, and nanoseconds and atomic clocks or even TV schedules. Yet there are links as well, for time has long been a factor in human and personal organization, whether in relation to the seasons or in dealings with other human beings or even with machines.

Certainly time, which started out as an interaction between human societies and nature, has had an active history. The process, even today, has involved complex combinations of objective factors, like the movement of sun and moon, or the speed of steam-driven machinery, and human ideas and impulses. Changes in time have reflected significant expressions of religious belief, plus the calculations of scientists and mathematicians, plus the interests of the state and the ruling class, plus the needs of urban environments, plus, of course, the available capacities for time measurement. The uses of time have translated many larger forces into the basic structures of daily life – literally, for entire populations at some points.

From the outset, as we have seen, complexities enter into the human calculations of time. Nature does not give consistent signals: the sun and the moon offer different systems, and even the natural rhythm of the human body, though it can be adjusted to the daily cycle of the sun, is just a bit different. Here is an ample basis both for variety, in the time systems humans have devised, and ongoing debates over what kind of time works best. Without question, human invention is essential, and ultimately many of the most important calculations of time are simply human conventions, with only limited reference to nature. "I'll be there in fifteen minutes" is a human construct.

The varied history of time certainly shows how humans can adapt to an array of different timing systems, depending on the belief patterns and economic arrangements in play at any given point. A favorite illustration might be the long-held notion that the hours of the day should vary in length, depending on the season. In many periods also, people's capacity to juggle different time systems, for different purposes, is also a tribute both to human flexibility and to the various uses for constructions of time. Thanks to globalization, this variety is diminishing somewhat, but it is still visible in many regions of the world. Most people still use a number of time systems depending on the activity engaged.

Time's history also reminds us that some of the annoyances societies are still trying to deal with today – like the often intense debate over daylight savings – reflect problems that people have been addressing at least since the advent of the sundial. If nothing else, it remains difficult to accommodate the fact that sunlight alters with the seasons.

At many junctures, and not just in the most recent periods of history, time has reflected the complexities of change. Different levels of need and knowledge have caused variations in the ways major groups, even in the same society, have interpreted time. Many innovations took hold more slowly, and less completely, than might be imagined given the factors involved. While many people have had new conceptions of time imposed on them, they have often been able to resist and modify in ways that qualified the ultimate outcomes.

Complexity has also marked the intersections between time and globalization, particularly from the 18th century onward. Time can be used to illustrate the power of global contacts and the need to adjust basic arrangements to the demands of more rapid and extensive exchanges among the major regions of the world. For many decades, time and imperialism seemed to go hand in hand. But the result, at least to date, has been some common global trends but not a single global culture of time. Time systems have been seriously adjusted, over the past 200 years, to facilitate global contacts, but there is regional as well as personal flexibility alongside the shared patterns. Global history and more local histories combine, and comparative analysis remains essential.

The comparative history of time has also raised important questions about the impact and significance of different patterns of adjustment to change. Was the approach to time an important factor in the shifting relationship between the West and many other parts of the world? Did it reflect or cause other distinctive Western patterns? And how has the comparative picture altered, particularly during the past 200 years? Are particular regional approaches to time still significant factors in economic and cultural distinctions, as for example some of the "pace of life" analyses suggest?

Ultimately the overarching question about the history of time involves an evaluation of the distortions time has imposed, and continues to impose, on human life. For various reasons, and not just the relevant blend of economics and technology, human cultures increasingly crafted units of time beyond the incorporation of signals from nature. Most people today, at least in the cities, devote more attention to knowing the hour than in noting sunsets, though the latter are not irrelevant; and most people think more about weekends than solstices.

We have seen that complaints about the burdensome aspects of human time begin early, well before people were exposed to the rhythms of mechanical clocks. Clearly, pressures increased as new ideas and timekeeping technologies combined. It is not hard to paint a picture of the modern human – including the modern child at least after infancy – dominated by time-based schedules that have nothing to do with individual impulse or variety, that squeeze out true play and spontaneity. Without question, for some people, the modern history of time contributes actively to a modern history of personal anxiety and certain physical disorders, and even repression. We have also seen that, at least in modern societies, time sense can be translated into nasty prejudices, against individuals and even groups whose punctuality does not measure up: the link between time and moralism can be powerful, and that remains true today.

But judgments should be complicated, for several reasons. First, we have seen that people can acclimate to various new systems of time, though the adjustments may take a while. Once people are used to the clock, raised by it, it may not prove entirely burdensome. Indeed – and this has been a key use of time from early societies onward – time helps provide a sense of order on what might otherwise seem a chaotic environment.

Further, even aside from personal variations and different regional cultures, the human capacity for dealing with different times systems for different purposes deserves attention in the present as well as the past. Hourly timing may have crept into the weekend, but it does not always predominate. Leisure has succumbed to timing to some extent, but modern developments have also tossed up some escape valves. One humble example: thanks first to videocassette recorders (VCRs) and now to digital recording, many people are no longer tied to particular times for their television viewing, and indeed the growing popularity of binge watching is a deliberate defiance of normal time pressures.

Thinking about the complexities of modern time offers a way to assess the quality of many aspects of modern life, and even to take a look at our own, personal approaches to time – but it does not yield easy judgments.

A key issue, from a world history standpoint, involves not just comparison but comparative evaluation of to which extent the Western (or Japanese) time culture, now adopted in other parts of the world, is a desirable or even essential framework for modern life. It certainly provides bases for criticism: punctuality is often easy to measure and the history of time helps explain why it often carries moral qualities as well. But many people, in the West and elsewhere, have found time burdensome, and have sought to carve out lives less constrained by time. Is there a best approach in these matters, or must we simply grant the validity of different cultural approaches – let the punctual Dutch hosts do the best they can with the tardy Indian visitors (and vice versa); they can probably work things out – and leave it at that? Will it prove essential, in a globalized world, to do more to insist on a more uniform basic standard?

Take, for example, the virtual meeting, an increasingly common phenomenon in the 21at century. Over the internet, interlocutors are gathered from Scandinavia, Australia, China, plus two time zones in the United States. Some people are calling in early in the morning, some late in the evening. But, thanks to modern videoconferencing, the package works, after elaborate scheduling: but everyone must be on time or the effort collapses. Hard to argue that cultural variety works in this setting.

But against this, the argument that different societies bring different values into the equation, that there is no one size that fits all –again, apart from differences in personal commitment to punctuality – raises important questions about whether the Western approaches to global demands about time arrangements are culturally insensitive and possibly counterproductive. The long history of prejudice against some time cultures does not require enthusiastic embrace of every past system, in societies that require better coordination, but it certainly urges caution. Maximum efficiency may not always be the top goal.

And what of the future? Some aspects of life will surely become subject to even more precise timing. Robots, most obviously, can be timed in ways that humans cannot, and many people may have to adjust schedules accordingly. But growing use of robots, whatever the overall impact, may also free humans from some of the most rigidly timed tasks. Global time issues may generate further change. A few visionaries have argued that the whole world should now operate by a single clock, given the undeniable nuisance of having to take into account the different time zones. The argument urges that being able to communicate around the world at the same time, given the possibilities of the Internet, will become more important than having every region able to start the day near dawn. After all, the two huge countries that already have single time zones, India and China, routinely ask their populations to operate in quite varied relationships to the daily paths of the sun.

But regional preferences concerning time will surely not disappear, given the link between particular notions of time and various national and religious identities. While robot time may become more important, increasing attention to health needs and growing population segments like the elderly may provide some counterbalance, requiring new flexibilities where time is concerned. Other reforms, widely discussed, might seek to reduce the burdens of rigid scheduling on children.

The history of time, like the history of most subjects, does not lend itself to precise forecasts. But the history does highlight some of the key issues involved, and some of the most powerful recent trends. It certainly promotes discussion of the differences and overlaps between modern people and their earlier counterparts, where time is concerned. Furthermore, it at least helps frame a discussion of how time might change in future – and how we should want it to change. And in a globalization context, as in the imperialist context of the 19th century, it raises real questions of tolerance. How widely can the most rigid time sense, particularly acute in the West, be used to judge the experience of other societies? What is the healthiest, and not just most productive, approach to time in any case?

Finally, the history of time has been wrapped up in the history of patience, though this is a tough topic historically. By modern standards lots of situations in the past, where arrival times were hard to agree upon or people simply showed up for a gathering in dribs and drabs, must have required great calm to endure – though a more relaxed sense of time, not tied to the clock, might have mitigated considerably. Not surprisingly, most major religions had placed a premium on patience as a virtue. In contrast, modern time systems place fewer burdens on patience; if we all show up within five minutes and at the same time are culturally conditioned to apologize sincerely for even moderate tardiness, there is not much to worry about. But these same systems raise expectations, and minor deviations can easily seem unacceptable: perhaps not five minutes late, but ten? Other societies that have shared a modern evolution toward time but less rigorously than the West may have a different take on patience, with less stress resulting. Time, and changes in time systems, raise big questions about the quality of life, and they may not have easy answers.

There is no doubt that many modern people are highly timed, and certainly different from their ancestors. The key issues are, first, whether they suffer from rigidity or, trained to the clock from childhood, enjoy greater predictability; and, second, how they react, in a globalizing environment, to people or groups who march to a slightly different drummer. The modern history of time may be a story of growing constraint and imposition; or an increase in efficiency and order; or a complex combination of both.

We may have passed through the roughest segments in the modern history of time: the confrontation with novelty was much greater in the early modern centuries, in the West, and in the 19th/early 20th centuries in Western interactions with other parts of the world. But it is fair to say that many of us are not entirely comfortable about time – just think of the worries about allocating hours to sleep, or the anxious discussions about whether we are overscheduling our children; and cultural differences in time habits complicate the picture still further. It is valid, moving out from a historical understanding of how time has evolved, to talk about ways to loosen the hold of time. It is valid, looking at time's history over the past four or five centuries, to ask whether time has seized us, rather than the other way around.

Index